Step by Step
ELECTRICAL ENGINEERING

No Background Required

authored by

Gregor Haynes

Copyrighted Material

Copyright © 2024 Gregor Haynes. All rights reserved.

No part of this book may be reproduced, distributed, transmitted, or stored in any form or by any means, including photocopying, recording, or other electronic or mechanical methods, without the prior written permission of the author, except as permitted by copyright law.

This book is intended for personal use only. Any sale, distribution, or sharing of the content with third parties without written authorization is strictly prohibited. Legal action will be pursued for any violations.

Disclaimer:

The information contained in this book is provided to the best of the author's knowledge and expertise and is intended for educational and informational purposes only. No guarantees of any kind are made regarding the accuracy, completeness, or adequacy of the material for specific situations. Readers are encouraged to seek professional advice for their particular circumstances.

The author assumes no responsibility for any loss, damage, or inconvenience caused directly or indirectly by the use of the information contained in this book, including errors, omissions, or misinterpretations. Readers are solely responsible for their use of the information and any results obtained from it.

Legal Notice:

This book does not constitute legal, financial, engineering, or other professional advice. Any attempt to replicate the information contained in this book is done at the reader's own risk and responsibility. The author makes no guarantees that the skills or outcomes described in the book are achievable by all readers.

Table of Contents

Who This Book Is For VII

Preface VIII

    Introduction to Electrical Engineering VIII

    Purpose of the Book IX

    How to Use This Book X

Chapter 1 1

The History and Importance of Electricity 1

    Introduction 1

    1.1.   Electricity in Nature 1

    1.2.   The History of the Discovery of Electricity 2

    1.3.   Methods of Generating Electricity 3

    1.4.   Implications of a World Without Electricity 5

Chapter 2 7

Fundamentals of Electricity 7

    Introduction to Electricity 7

    2.1.   Electric Charge and Current 7

    2.2.   Voltage and Potential Difference 11

    2.3.   Grounding and Its Importance in Electrical Engineering 17

    2.4.   Basic Concepts of Resistance and Conductance 20

    2.5.   Measuring Resistance and Conductance 22

    2.6.   Quiz for Review 25

    2.7.   Practical Exercises 26

    2.8.   Curiosities and Further Insights 27

    References and Further Reading for Chapter 2 30

Chapter 3 31
Simple Electrical Circuits 31

    3.1.   Basic Components: Resistors, Capacitors, Inductors 31

    3.2.   Series and Parallel Circuits 36

    3.3.   Ohm's Law and Kirchhoff's Laws 41

    3.4.   Quiz for Review 47

    3.5.   Practical Exercises 48

    3.6.   Curiosities and Further Insights 49

    References and Further Reading for Chapter 3 53

Chapter 4 54
Magnetism and Electromagnetism 54

    Introduction to Magnetism 54

    4.1.   Magnetic Fields Generated by Current 58

    4.2.   Electromagnets and Their Applications 61

    4.3.   Quiz for Review 66

    4.4.   Practical Exercises 67

    4.5.   Curiosities and Further Insights 68

    References and Further Reading for Chapter 4 71

Chapter 5 72
Alternating Current (AC) and Direct Current (DC) 72

    5.1.   Differences Between AC and DC 72

    5.2.   Uses of AC and DC in Everyday Applications 75

    5.3.   Introduction to Transformers 79

    5.4.   Quiz for Review 83

    5.5.   Practical Exercises 84

    5.6.   Curiosities and Further Insights 85

References and Further Reading for Chapter 5 89

Chapter 6 90

Semiconductors and Electronic Components 90

Introduction to Semiconductors 90

6.1. Diodes, Transistors, and Their Uses 94

6.2. Advanced Overview of Essential Electronic Components ... 98

6.3. Quiz for Review 102

6.4. Practical Exercises 103

6.5. Curiosities and Further Insights 104

References and Further Reading for Chapter 6 107

Chapter 7 108

Practical Applications and Projects in Electrical Engineering .. 108

Introduction to Practical Applications 108

7.1. Consumer Electronics 112

7.2. Home Automation and IoT (Internet of Things) 113

7.3. Renewable Energy Systems 116

7.4. Security and Monitoring Systems 118

7.5. Lighting Systems 121

7.6. Robotics and Control Systems 125

7.7. Mobile and Portable Power Solutions 129

7.8. Curiosities and Further Insights 133

Chapter 8 136

Safety and Maintenance 136

8.1. Principles of Electrical Safety 136

8.2. Basic Maintenance of Electrical Systems 139

8.3. Guidelines for Home and Workplace Safety 143

Access Your High-Resolution PDF Version 148
    Optimized Formatting and Easy Reference 148

Appendices 149
    Glossary of Technical Terms 149
    Reference Tables and Constants.. 155
    Electronic Circuit Symbols.. 159
    Resistor Color Code (5 Bands) 161

Index.. 162

Who This Book Is For

This book is your gateway to mastering the essentials of electrical engineering, crafted for anyone who aspires to understand and apply the principles of this dynamic field.

Whether you're a student, a hobbyist, or someone with a keen interest in the electrical systems that power our world, this book is designed to meet you where you are.

Unlike other books that promise to make you an electrical engineer in just seven days—a feat that is unrealistic—this book offers a solid foundation, equipping you with the knowledge and practical skills necessary to truly comprehend and engage with electrical engineering.

With clear explanations, practical exercises, and insights into real-world applications, this book will set you on the path to becoming proficient in electrical engineering, preparing you for further study or simply satisfying your curiosity.

If you're ready to embark on a journey of discovery and build a strong understanding of electrical engineering, **this book is for you.**

Preface

Introduction to Electrical Engineering

Electrical engineering is one of the most dynamic and impactful fields in modern technology, influencing nearly every aspect of our daily lives. From the power systems that light our homes to the complex electronics in our smartphones, electrical engineering is at the heart of the devices and systems that define the modern world.

At its core, electrical engineering is the study and application of electricity, electronics, and electromagnetism. It encompasses a wide range of topics, including power generation and distribution, signal processing, telecommunications, and control systems. Electrical engineers design, develop, and maintain electrical equipment and systems, ensuring that they operate efficiently, reliably, and safely.

The field has a rich history, with roots stretching back to the discovery of electricity itself. Pioneers such as Michael Faraday, James Clerk Maxwell, and Nikola Tesla laid the foundations for the technologies we use today. Over the decades, electrical engineering has evolved, integrating new technologies and adapting to the changing needs of society. Today, it plays a crucial role in addressing global challenges such as renewable energy, smart infrastructure, and sustainable development.

As we move further into the 21st century, the importance of electrical engineering continues to grow. Innovations in areas like renewable energy, automation, and the Internet of Things (IoT) are opening new frontiers and creating opportunities for engineers to make significant contributions to the world.

This book is designed to introduce you to the fundamentals of electrical engineering, offering a practical guide to help you understand and apply key concepts. If you're new to the field or looking to refresh your knowledge, this book aims to provide a solid foundation that will prepare you for further study and exploration in electrical engineering.

Electrical engineering is not just about theory; it's about solving real-world problems and making tangible improvements to the systems that power our lives. By the end of this book, you will have gained a deeper understanding of the field and be equipped with the knowledge and skills needed to embark on your own journey in electrical engineering.

Purpose of the Book

The purpose of this book is to provide a comprehensive and accessible introduction to the field of electrical engineering. Recognizing that electrical engineering can often seem complex and intimidating, this book is designed to break down the essential concepts into manageable, understandable parts, making the subject approachable for beginners and those without extensive backgrounds in mathematics and physics.

This book serves several key purposes:

Foundational Knowledge: It aims to equip readers with the foundational knowledge necessary to understand and work with electrical systems. This includes basic concepts such as voltage, current, and resistance, as well as more advanced topics like semiconductors, circuit design, and renewable energy systems.

Practical Application: Beyond theoretical knowledge, this book emphasizes the practical application of electrical engineering principles. Through hands-on projects, exercises, and real-world examples, readers will learn how to apply what they've learned to solve problems and create functional electrical systems.

Preparation for Further Study: For those interested in pursuing a career or further studies in electrical engineering, this book serves as a stepping stone. It provides the foundational understanding needed to advance to more specialized or complex topics in the field.

Fostering Interest and Curiosity: Electrical engineering is a field filled with innovation and discovery. This book is designed not just to educate, but to inspire curiosity and interest in the possibilities that electrical engineering offers. By presenting the material in an engaging and accessible way, it aims to spark enthusiasm and encourage readers to explore further.

Safety and Best Practices: A strong emphasis is placed on safety and best practices throughout the book. Electrical engineering involves working with potentially hazardous systems, and understanding how to do so safely is paramount. This book provides guidelines and practical advice to help readers develop safe habits and practices from the very beginning.

In summary, the purpose of this book is to serve as a gateway into the world of electrical engineering, even for those with no prior knowledge.

How to Use This Book

This book is structured to provide a clear and progressive learning experience in electrical engineering. Each chapter builds on the concepts introduced in the previous ones, gradually increasing in complexity while remaining accessible. To make the most of this book, here are some guidelines on how to use it effectively:

Start with the Basics: If you're new to electrical engineering, it's important to start at the beginning and work your way through the chapters in order. The early chapters lay the groundwork for understanding more advanced topics later in the book. Don't rush through the basics—having a solid foundation is key to mastering the more complex concepts.

Engage with the Practical Exercises: This book is not just about reading; it's about doing. Each chapter includes practical exercises and projects that are designed to reinforce the concepts you've learned. These hands-on activities are crucial for deepening your understanding and developing practical skills. Be sure to take the time to complete these exercises, as they will help solidify your knowledge.

Use the Quizzes for Self-Assessment: At the end of each chapter, you'll find a quiz designed to test your understanding of the material. These quizzes are a great way to review what you've learned and identify any areas where you might need further study. Take these quizzes seriously, and if you find certain topics challenging, don't hesitate to revisit the relevant sections.

Refer to the Curiosities and Insights: Each chapter includes a section on curiosities and further insights that delve into interesting facts, historical anecdotes, and advanced applications of the concepts covered. These sections are meant to inspire and expand your understanding of how electrical engineering impacts the world. They also offer a glimpse into more advanced or specialized topics that you might explore in the future.

Consult the Bibliography and Further Reading: At the end of the book, you'll find a bibliography and a list of recommended further reading. These resources are invaluable if you wish to delve deeper into specific topics or continue your education in electrical engineering. The suggested readings offer both theoretical knowledge and practical insights from experts in the field.

Maintain a Learning Journal: As you work through the book, consider keeping a learning journal. Use it to take notes on key concepts, jot down questions or ideas, and track your progress through the exercises and quizzes. A journal can be a powerful tool for reinforcing your learning and reflecting on your progress.

Stay Safe: Electrical engineering involves working with potentially dangerous systems. Throughout the book, safety guidelines are provided to help you avoid accidents and injuries. Always follow these guidelines carefully, especially when working on the practical projects. Safety should always be your top priority.

Pace Yourself: Learning electrical engineering is a journey, and it's important to pace yourself. Take the time to thoroughly understand each concept before moving on to the next. Don't be discouraged if some topics take longer to grasp—everyone learns at their own pace.

By following these guidelines, you'll be well-equipped to navigate the material in this book and gain a solid understanding of electrical engineering. Whether your goal is to pursue a career in the field, enhance your existing knowledge, or simply satisfy your curiosity, this book is designed to support your journey every step of the way.

Chapter 1

The History and Importance of Electricity

Introduction

Electricity is a cornerstone of modern civilization, integral to virtually every aspect of our daily lives. From the moment we wake up to the hum of our alarm clocks to the lights that illuminate our nights, electricity powers our world. Understanding its origins, its presence in nature, and the methods by which we harness it is essential for anyone delving into electrical engineering. This chapter will take you through the fascinating journey of electricity, from its natural manifestations to its historical discoveries and its critical role in contemporary society.

1.1. Electricity in Nature

Natural Electrical Phenomena

Electricity is not a human invention; it exists naturally in the environment. One of the most dramatic displays of natural electricity is lightning, a powerful discharge of static electricity that occurs during thunderstorms. Lightning can reach temperatures of around 30,000 Kelvin, much hotter than the surface of the sun, and release immense amounts of energy in a fraction of a second.
This phenomenon not only captivates us with its visual splendor but also underscores the raw power and energy inherent in electrical forces.
Beyond lightning, many plants and animals exhibit electrical phenomena. For instance, the Venus flytrap, a carnivorous plant, uses electrical signals to trigger its rapid closing mechanism, trapping unsuspecting insects. Similarly, certain fish, such as the electric eel, generate electric fields to navigate murky waters, communicate with each other, and incapacitate prey. These biological marvels illustrate the diverse and vital roles electricity plays in the natural world.

Early Observations

The first recorded observations of electrical phenomena date back to ancient Greece. Around 600 BCE, Thales of Miletus noted that rubbing amber with fur would attract small objects like feathers. This simple experiment was an early glimpse into the world of electrostatics. The term "electricity" itself is derived from the Greek word "elektron", meaning amber. This connection highlights how the ancients' curiosity about natural phenomena laid the groundwork for future scientific exploration.

1.2. The History of the Discovery of Electricity

The journey to our modern understanding of electricity is rich with discoveries and innovations that span centuries. Early observations laid the groundwork for the more sophisticated explorations that followed.

17th and 18th Centuries

The exploration of electricity advanced significantly in the 17th and 18th centuries. Benjamin Franklin, with his famous kite experiment in 1752, demonstrated that lightning was a form of electrical discharge. By flying a kite during a thunderstorm and observing sparks jumping from a key attached to the kite string, Franklin provided evidence that lightning and electricity were the same phenomena. This was a pivotal moment in understanding the connection between natural and artificial electricity and marked a significant step in the study of electrical science.

Alessandro Volta's invention of the voltaic pile in 1800 marked the creation of the first true battery, providing a steady source of electrical current. The voltaic pile, composed of alternating discs of zinc and copper separated by brine-soaked cardboard, demonstrated that chemical reactions could produce electricity. Volta's work laid the foundation for future electrical experiments and applications, earning him the honor of having the unit of electric potential, the volt, named after him.

Michael Faraday's discovery of electromagnetic induction in 1831 was another monumental breakthrough. Faraday demonstrated that moving a conductor through a magnetic field could generate an electric current.

This principle, known as Faraday's Law of Induction, underpins modern electrical generators and transformers. Faraday's experiments with coils of wire and magnets showed that a changing magnetic field could induce an electric current, revolutionizing our ability to generate and manipulate electricity.

19th Century

The 19th century was a period of rapid advancement in electrical technology. Nikola Tesla's development of alternating current (AC) systems allowed electricity to be transmitted over long distances more efficiently than direct current (DC) systems, which were championed by Thomas Edison. This "War of Currents" ultimately saw AC become the dominant method of power distribution. Tesla's inventions, including the AC induction motor and the Tesla coil, paved the way for widespread adoption of AC power, which is still the standard for electrical transmission today.

Thomas Edison, meanwhile, was instrumental in creating the first practical electric light bulb and establishing the first power stations. Edison's development of the incandescent light bulb in 1879 revolutionized indoor lighting, making it more reliable and accessible. He also pioneered the creation of power generation and distribution systems, setting up the first electric power station in New York City in 1882. Edison's contributions to electrical engineering and his business acumen helped accelerate the adoption of electricity in homes and industries.

1.3. Methods of Generating Electricity

Electricity generation has evolved from simple mechanical processes to sophisticated technologies harnessing various energy sources.

Traditional Methods

Traditional methods of electricity generation include burning fossil fuels like coal and gas. These fuels are combusted to heat water, producing steam that drives turbines connected to generators. This process, known as thermal power generation, has been a cornerstone of electricity production for over a century. However, it also poses significant environmental challenges due to greenhouse gas emissions and air pollution.

Hydroelectric power, another traditional method, utilizes the energy of falling water to turn turbines. This method, often employed in large dams, converts the kinetic energy of water into mechanical energy and then into electrical energy. Hydropower is a renewable source of energy and can produce significant amounts of electricity without emitting greenhouse gases, making it an essential component of sustainable energy strategies.

Modern Methods

Modern methods have expanded to include nuclear power, which uses nuclear reactions to produce heat for generating electricity. Nuclear power plants rely on the process of nuclear fission, where atomic nuclei are split to release energy. This heat is used to produce steam that drives turbines. Nuclear power is capable of generating large amounts of electricity with relatively low greenhouse gas emissions, but it also raises concerns about radioactive waste and the potential for catastrophic accidents.

Renewable energy sources are also becoming increasingly important. Solar power converts sunlight into electricity using photovoltaic cells, while wind power harnesses the kinetic energy of wind using turbines. Geothermal energy taps into the Earth's internal heat, and biomass energy derives from organic materials. These renewable sources are essential for reducing dependence on fossil fuels and mitigating climate change. Innovations in solar panel efficiency, wind turbine design, and energy storage technologies are continually enhancing the viability and scalability of renewable energy.

Future Technologies

Looking forward, fusion power holds the promise of nearly limitless energy by mimicking the processes that power the sun. Fusion involves combining light atomic nuclei to form a heavier nucleus, releasing enormous amounts of energy. Unlike nuclear fission, fusion produces minimal radioactive waste and carries a lower risk of accidents. Research into fusion technology, such as the International Thermonuclear Experimental Reactor (ITER) project, aims to make fusion a practical and sustainable energy source in the future.

Additionally, advancements in energy storage technologies, such as advanced batteries, are critical for maximizing the efficiency and reliability of renewable energy sources. Effective energy storage solutions can help balance supply and demand, ensuring a steady and reliable electricity supply even when renewable sources are intermittent. Innovations in battery chemistry, such as lithium-sulfur and solid-state batteries, promise higher energy densities and longer lifespans, further enhancing the feasibility of renewable energy integration.

1.4. Implications of a World Without Electricity

Imagining a world without electricity highlights its vital role in our lives and the potential consequences of its absence.

Impact on Daily Life

Without electricity, modern communication systems would collapse. Smartphones, computers, and the internet rely entirely on a stable power supply. Transportation systems, including electric trains and traffic lights, would grind to a halt, causing widespread disruption. Air travel would be severely impacted, as airports and air traffic control systems depend on reliable electricity.

In agriculture, the production, storage, and transportation of food are heavily dependent on electrical systems. Refrigeration units, irrigation systems, and processing facilities all require electricity to function. A lack of electricity would severely impact food security and availability, leading to potential food shortages and increased prices.

Healthcare systems would face catastrophic challenges without electricity. Hospitals rely on electrical power for lighting, medical equipment, and life-support systems. Diagnostic tools such as MRI and CT scanners, as well as surgical instruments, depend on electricity. In the absence of reliable power, healthcare delivery would be compromised, endangering lives.

Impact on Economy and Society

The economic impact of a world without electricity would be catastrophic. Industries and businesses rely on electricity for production, operations, and communication.

Manufacturing plants would cease to operate, leading to job losses and economic decline. Financial institutions and stock markets would be unable to function, causing economic instability.

A prolonged absence of electricity could lead to massive economic losses and societal regression, undoing centuries of progress. The global economy is intricately linked to the availability of reliable electricity, and disruptions could trigger a cascade of failures across multiple sectors. Social services, including education, would be heavily impacted, as schools and universities depend on electricity for lighting, computing, and communication.

Reflecting on the history and methods of generating electricity helps us appreciate its crucial role in modern society. Understanding how electricity is harnessed and utilized today provides a foundation for exploring its applications and innovations in electrical engineering. As we move forward, the continued development of sustainable and efficient energy sources will be essential for maintaining and enhancing our way of life.

Chapter 2

Fundamentals of Electricity

Introduction to Electricity

Electricity is a fundamental aspect of the physical world, crucial not only to our everyday lives but also to a wide range of scientific and engineering disciplines. To understand electricity, we must delve into the behavior of electric charges, the forces they exert, and the ways in which these forces can be harnessed to perform useful work.

2.1. Electric Charge and Current

Electric Charge

Electric charge is a fundamental property of matter, integral to the behavior of subatomic particles. It is the source of the electromagnetic force, one of the four fundamental forces of nature. Electric charge comes in two types: positive and negative. Protons carry a positive charge, while electrons carry a negative charge. Neutrons, on the other hand, are electrically neutral, meaning they have no charge.

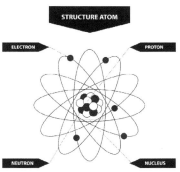

Fundamentals of Electricity

The unit of electric charge in the International System of Units (SI) is the coulomb (C). One coulomb is equivalent to approximately 6.242×10^{18} elementary charges (the charge of a single proton or electron). The charge of a single electron is approximately -1.602×10^{-19} coulombs, while the charge of a proton is $+1.602 \times 10^{-19}$ coulombs.

Properties of Electric Charge

Quantization: Electric charge is quantized, meaning it exists in discrete units rather than a continuous spectrum. The smallest unit of charge is the elementary charge, which is the charge of a single proton or electron. This principle of quantization implies that the charge on any object is an integer multiple of the elementary charge.

Conservation: The principle of conservation of charge states that the total electric charge in an isolated system remains constant over time. This means that charge can neither be created nor destroyed but can only be transferred from one object to another.

Attraction and Repulsion: Like charges repel each other, while opposite charges attract. This behavior is described by Coulomb's law, which quantifies the electrostatic force between two point charges.

Coulomb's Law

Coulomb's law provides a quantitative description of the force between two charges. It states that the magnitude of the electrostatic force F between two point charges q_1 and q_2 is directly proportional to the product of the magnitudes of the charges and inversely proportional to the square of the distance r between them:

$$F = k_e \frac{q_1 q_2}{r^2}$$

where k_e is Coulomb's constant ($8.9875 \times 10^9 \, \text{N·m}^2/\text{C}^2$).

This force acts along the line joining the two charges and is attractive if the charges are of opposite sign and repulsive if the charges are of the same sign.

Electric Current

Electric current is the flow of electric charge. In most practical applications, this flow is due to the movement of electrons in a conductor, such as a metal wire. The direction of the conventional current is defined as the direction in which positive charges would move, which is opposite to the actual flow of electrons.

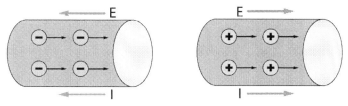

The unit of electric current is the ampere (A), which is defined as one coulomb of charge passing through a given point in a circuit per second:

$$I = \frac{Q}{t}$$

where I is the current, Q is the charge, and t is the time.

Types of Current

Direct Current (DC): In direct current, the flow of electric charge is unidirectional. DC is commonly used in battery-powered devices and electronic circuits. The current flows from the positive terminal to the negative terminal of the power source.

Alternating Current (AC): In alternating current, the flow of electric charge periodically reverses direction. AC is the standard form of electricity delivered to homes and businesses. The current changes direction in a sinusoidal pattern, with the voltage oscillating between positive and negative values.

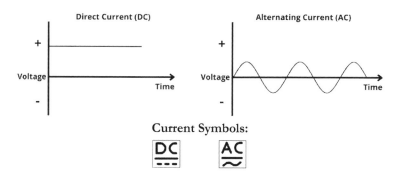

Measuring Electric Current

Electric current can be measured using an instrument called an ammeter. An ammeter is connected in series with the circuit so that the current flows through the meter. This allows the ammeter to measure the total current flowing through the circuit. For accurate measurements, the ammeter must have a very low internal resistance to minimize its impact on the circuit.

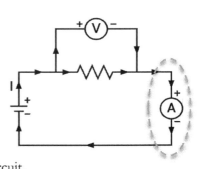

Applications of Electric Current

Electric current is fundamental to countless applications in modern life. Some of the key applications include:

Powering Electrical Devices: Electric current is essential for powering a wide range of devices, from household appliances to industrial machinery. The flow of current provides the energy needed to operate these devices.

Communication Systems: Current flows in communication systems, enabling the transmission of signals in telephones, radios, and televisions. These signals can carry voice, video, and data over long distances.

Medical Equipment: Many medical devices rely on electric current for their operation. Examples include diagnostic tools such as electrocardiograms (ECGs), which measure the electrical activity of the heart, and imaging devices like MRI machines.

Transportation: Electric current is used in various modes of transportation, including electric cars, trains, and airplanes. The development of electric vehicles is a significant step toward reducing carbon emissions and reliance on fossil fuels.

Conductors, Insulators, and Semiconductors

Materials are classified based on their ability to conduct electric charge.

Conductors: Conductors, such as metals (e.g., copper, aluminum), have a high density of free electrons that can move easily, allowing electric current to flow with minimal resistance.

Insulators: Insulators, such as rubber, glass, and plastic, have tightly bound electrons that do not move freely, preventing the flow of electric current. These materials are used to protect and insulate conductors in electrical circuits.

Semiconductors: Semiconductors, such as silicon and germanium, have properties between those of conductors and insulators. They are essential in modern electronics, as they can be manipulated to conduct or insulate under different conditions. Semiconductors form the basis of devices like diodes, transistors, and integrated circuits.

Transitioning from understanding electric charge and current, let's explore how these concepts lead to the creation of voltage and potential differences in electrical circuits.

2.2. Voltage and Potential Difference

Introduction to Voltage

Voltage, also known as electric potential difference, is a measure of the potential energy per unit charge between two points in an electric field. It is the force that drives electric current through a circuit, analogous to the pressure that pushes water through a pipe. The unit of voltage is the volt (V), which is equivalent to one joule per coulomb (J/C).

Understanding Electric Potential

Electric potential at a point in an electric field is defined as the work done in bringing a unit positive charge from infinity to that point without any acceleration. When we talk about voltage, we often refer to the potential difference between two points. This potential difference is what causes charges to move in a circuit.

Fundamentals of Electricity

Imagine a charged particle in an electric field. If the particle moves from a point of higher potential to a point of lower potential, work is done by the electric field on the particle. Conversely, to move a charge against the electric field, work must be done on the charge. The amount of work done per unit charge is the potential difference or voltage.

Electric Field (E)

Definition: The electric field is a vector field that represents the force a charge would experience per unit charge at any point in space. It is a fundamental concept in understanding how charges interact with each other in an electric field.

- **Formula:** The electric field E at a point is defined by the equation:

$$E = \frac{F}{Q}$$

where E is the electric field, F is the force experienced by the charge, and Q is the magnitude of the charge.

Relationship with Potential Difference: The electric field is closely related to the electric potential difference (voltage) between two points in space. The potential difference between two points is the work done per unit charge to move the charge between those points in the presence of an electric field.

- **Formula:** For a uniform electric field, the potential difference V between two points separated by a distance d is given by:

$$V = E \cdot d$$

where V is the potential difference, E is the magnitude of the electric field, and d is the distance between the two points.

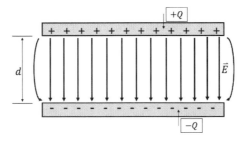

Direction of the Electric Field: The direction of the electric field is defined as the direction in which a positive test charge would be pushed or pulled. The field points away from positive charges and toward negative charges. The strength of the field indicates the magnitude of the force that a charge would experience.

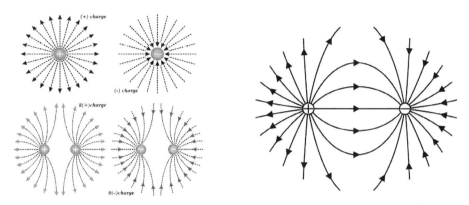

Applications: Understanding the electric field is essential for analyzing circuits, designing electronic components, and understanding phenomena such as electrostatic induction. The concept of the electric field also plays a crucial role in explaining how voltage is established in different parts of a circuit.

Mathematical Representation

The potential difference V between two points is given by:

$$V = \frac{W}{Q}$$

where W is the work done in joules (J) and Q is the charge in coulombs (C).

In practical terms, if 1 joule of work is done to move 1 coulomb of charge between two points, the potential difference between those points is 1 volt.

Sources of Voltage

Voltage can be generated in various ways, each with its unique applications and characteristics.

Batteries: Batteries convert chemical energy into electrical energy. They consist of electrochemical cells with electrodes and electrolytes that facilitate chemical reactions, producing a flow of electrons and creating a potential difference between the terminals.

Dry Cell Battery

Generators: Generators convert mechanical energy into electrical energy using electromagnetic induction. When a conductor moves through a magnetic field, it induces an electric current, creating a voltage. This principle is utilized in power plants to generate electricity on a large scale.

Photovoltaic Cells: Solar cells convert light energy directly into electrical energy through the photovoltaic effect. When light photons strike the semiconductor material in the cell, they excite electrons, generating a flow of current and creating a potential difference.

Thermocouples: Thermocouples generate voltage based on the temperature difference between two different metals joined at one end. This temperature difference creates a potential difference, which can be used to measure temperature or generate small amounts of electricity.

Measuring Voltage

Voltage is measured using a device called a voltmeter. A voltmeter is connected in parallel with the component or points between which the voltage is to be measured. This configuration ensures that the voltmeter does not significantly alter the circuit's current.

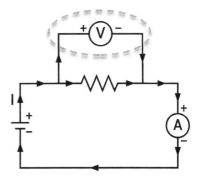

To measure the voltage across a battery, for example, the voltmeter is connected to the positive and negative terminals of the battery. The reading on the voltmeter indicates the potential difference between the terminals.

Applications of Voltage

Voltage is a critical parameter in many electrical applications and technologies. Here are a few key examples:

Power Supply: Voltage determines the amount of electrical power available to devices and systems. Different devices require different voltage levels to operate correctly, which is why power supplies are designed to provide the appropriate voltage for each application.

Signal Transmission: Voltage levels are used to encode information in communication systems. In digital electronics, binary information is represented by different voltage levels (e.g., 0 volts for a logic low and 5 volts for a logic high).

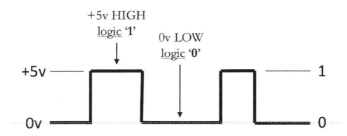

Electromotive Force (EMF): EMF is a type of voltage generated by changing magnetic fields, chemical reactions, or temperature differences. It is the driving force behind the generation of electrical energy in batteries, generators, and other devices.

Voltage Regulation: Many electronic devices include voltage regulators to maintain a constant voltage level, ensuring stable operation even when the input voltage fluctuates. Voltage regulation is crucial for the performance and longevity of electronic components.

Voltage in Everyday Life

Voltage plays a vital role in our daily lives, often without us even noticing it. Here are some common examples:

- **Household Appliances**: The standard voltage in household electrical outlets varies by country (e.g., 120V in the United States, 230V in Europe). This voltage powers a wide range of appliances, from refrigerators to televisions.

- **Batteries**: Batteries in devices like smartphones, laptops, and remote controls provide a convenient and portable source of voltage, enabling the operation of these devices without a direct connection to the power grid.

- **Automotive Systems**: Car batteries typically provide 12 volts of direct current (DC) to power the vehicle's electrical systems, including the starter motor, lights, and infotainment systems.

Safety Considerations

While voltage is essential for the functioning of electrical systems, it also poses safety risks. High voltages can cause electric shocks or burns if not handled properly. Here are some safety tips:

- **Insulation**: Ensure that all electrical connections are properly insulated to prevent accidental contact with live wires.

- **Proper Tools**: Use tools rated for the specific voltage level you are working with.

- **Awareness**: Be aware of the voltage levels in the systems you are working on, and take appropriate precautions to avoid hazards.

2.3. Grounding and Its Importance in Electrical Engineering

In electrical engineering, grounding (or earthing) refers to the practice of connecting part of a circuit to a reference point—often the Earth. Grounding provides a common return path for electric current and ensures safety, particularly in fault conditions. Proper grounding reduces the risk of electric shock, protects equipment, and ensures the stability of the electrical system.

Why Grounding is Critical

Grounding is crucial for several reasons:

Safety: By connecting electrical systems to the Earth, grounding provides a low-resistance path for fault currents. This reduces the risk of electrical shock if a short circuit occurs.

Overvoltage Protection: Grounding systems help dissipate excess voltage from events such as lightning strikes or power surges, protecting both people and equipment.

System Stability: Grounding helps maintain the stability of electrical systems by providing a consistent reference point for voltage levels. It ensures that electrical signals are correctly transmitted and received.

Electromagnetic Interference (EMI) Reduction: Grounding helps reduce noise and interference in circuits, particularly in sensitive systems such as communication or data networks.

Fundamentals of Electricity

Types of Grounding

There are several types of grounding systems, depending on the context and the application:

- **Earth Ground**: This is the most common form of grounding, where the system is physically connected to the Earth, usually via a metal rod driven into the soil. Earth grounding is essential for safety in electrical distribution systems.

- **Chassis Ground**: This is used to connect the metal casing of an electrical device to ground, providing protection in the event of a short circuit inside the device.

- **Signal Ground**: Signal grounding is primarily used in low-voltage systems, such as communication circuits, where the ground serves as a reference for the operation of the circuit. This helps prevent electrical noise from affecting the performance of the circuit.

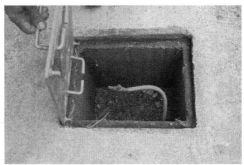

Fundamentals of Electricity

Grounding Symbols in Electrical Schematics

Grounding is represented by specific symbols in electrical schematics to identify different types of grounding connections. Understanding these symbols is crucial when analyzing or designing circuits.

Earth Ground: This symbol represents a direct connection to the Earth, typically for safety grounding in power distribution systems. It is used in designs where the circuit needs a physical connection to the Earth for fault protection.

Earth Ground

Example: This symbol is often used in high-power electrical systems to denote safety grounding.

Chassis Ground: This symbol denotes a connection to the metal frame or casing of a device. It protects users and equipment from electrical faults by ensuring that any stray current is conducted away safely.

Chassis Ground

Example: You'll see this symbol in electrical appliances where the metal case needs to be grounded for safety.

Signal Ground: Signal ground is used in low-voltage or sensitive electronic circuits to serve as the reference voltage. It provides a clean, noise-free return path for electronic signals, preventing interference.

Signal Ground

Example: Signal ground symbols are commonly used in audio and communication equipment.

Grounding Systems in Brief

In a typical grounding system, an electrical circuit's metal parts, casings, or equipment are connected to a grounding rod buried in the Earth. This connection ensures that in the event of a fault, such as a short circuit or insulation failure, the current will flow through the low-resistance path to the ground rather than through a person or sensitive equipment.

2.4. Basic Concepts of Resistance and Conductance

Introduction to Resistance and Conductance

Resistance and conductance are fundamental concepts in the study of electricity and are essential for understanding how electrical circuits function. Resistance is the property of a material that resists the flow of electric current, while conductance is the property that allows electric current to flow easily. These two properties are inversely related and play crucial roles in the design and operation of electrical circuits.

Resistance

Resistance, denoted by the symbol R, is measured in ohms (Ω). It quantifies how much a material opposes the flow of electric current. The resistance of a conductor depends on several factors:

Material: Different materials have different inherent resistivities. For example, metals like copper and aluminum have low resistivity and therefore low resistance, making them good conductors. Insulators like rubber and glass have high resistivity and high resistance. For a detailed comparison of the resistivity of various common materials, please refer to the "Resistivity of Common Materials" table at the end of this book.

Length: The longer the conductor, the higher the resistance. This is because the electrons encounter more obstacles as they travel through the material.

Cross-sectional Area: The larger the cross-sectional area of the conductor, the lower the resistance. A wider path allows more electrons to flow with less opposition.

Temperature: For most conductors, resistance increases with temperature. As temperature rises, the atoms in the conductor vibrate more, increasing the likelihood of collisions with the flowing electrons.

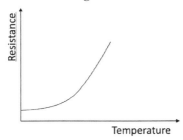

The relationship between resistance, resistivity (ρ), length (L), and cross-sectional area (A) of a conductor is given by the formula:

$$R = \rho \frac{L}{A}$$

Ohm's Law

A fundamental principle in the study of electrical circuits is Ohm's Law. It states that the current (I) flowing through a conductor between two points is directly proportional to the voltage (V) across the two points and inversely proportional to the resistance (R) of the conductor:

$$V = IR$$

Ohm's Law is crucial for analyzing simple circuits and understanding the relationship between voltage, current, and resistance. For more complex applications and circuits, see *Chapter 3: Simple Electrical Circuits*.

Conductance

Conductance, denoted by the symbol G, is measured in siemens (S). It is the reciprocal of resistance and indicates how easily electric current can flow through a material. The higher the conductance, the better the material conducts electricity. The relationship between conductance and resistance is given by:

$$G = \frac{1}{R}$$

2.5. Measuring Resistance and Conductance

Resistance can be measured using an instrument called an ohmmeter, which is often part of a multimeter. To measure resistance, the ohmmeter is connected across the component whose resistance is to be measured. For accurate measurements, the circuit should be powered off to prevent any current from affecting the reading.

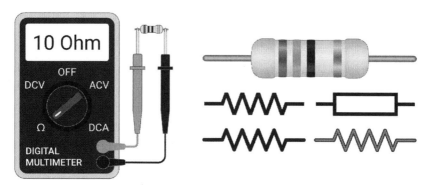

Conductance is less commonly measured directly but can be calculated if the resistance is known. High-conductance materials are used in applications where efficient current flow is critical, such as in wiring and electrical connections.

Applications of Resistance and Conductance

Understanding resistance and conductance is essential for designing and analyzing electrical circuits. Here are a few key applications:

Resistors: Resistors are components specifically designed to provide a certain amount of resistance in a circuit. They are used to control current flow, divide voltages, and protect sensitive components. For practical examples of using resistors in circuits, see *Chapter 3: Simple Electrical Circuits*.

Heating Elements: Devices like electric heaters and toasters use the resistance of a wire to generate heat. The electrical energy is converted into thermal energy as current flows through the high-resistance wire.

Sensors: Some sensors operate based on changes in resistance. For example, a thermistor changes its resistance with temperature, and a strain gauge changes its resistance with mechanical deformation.

Conductive Materials: Materials with high conductance, such as copper and aluminum, are used extensively in electrical wiring and components to ensure efficient current flow with minimal energy loss.

Practical Example: Designing a Voltage Divider

A voltage divider is a simple circuit that uses resistors to produce a specific fraction of the input voltage. It is often used in signal conditioning and sensor applications. The basic voltage divider circuit consists of two resistors in series connected to a voltage source. The output voltage (V_{out}) is taken from the junction of the two resistors.

The output voltage is given by:

$$V_{out} = V_{in} \left(\frac{R_2}{R_1 + R_2} \right)$$

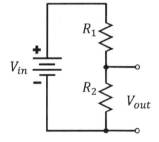

where V_{in} is the input voltage, and R_1 and R_2 are the resistances of the two resistors.

This principle is fundamental for creating reference voltages and scaling down signals in various electronic applications.

Fundamentals of Electricity

Safety Considerations

When working with electrical circuits, understanding resistance and conductance is crucial for ensuring safety. High resistance can lead to overheating, while low resistance can cause excessive current flow, potentially damaging components. Always use appropriately rated resistors and conductive materials to avoid hazards.

2.6. Quiz for Review

Access Additional Quiz Material

Scan the QR code below to download a PDF file with additional quiz questions. This allows you to practice offline and further reinforce your understanding of the concepts covered in the book.

2.7. Practical Exercises

Download Additional Exercises

Scan the QR code below to download a PDF file containing additional practical exercises. These exercises will help you apply the concepts learned in this book and further enhance your hands-on skills.

2.8. Curiosities and Further Insights

Exploring the world of electricity reveals fascinating phenomena and surprising applications. This section delves into some intriguing aspects of electricity that extend beyond basic principles, offering deeper insights and sparking curiosity.

The Triboelectric Effect: Static Electricity

One of the earliest observed electrical phenomena is static electricity, which results from the triboelectric effect. When certain materials come into contact and then separate, they can exchange electrons, leading to a buildup of electric charge. This is why you might experience a small shock after walking on a carpet and then touching a metal doorknob.

Historical Anecdote: In ancient Greece, Thales of Miletus discovered that rubbing amber with fur attracted small objects. This observation laid the groundwork for understanding electrostatics, as "electron" derives from the Greek word for amber.

Electric Fish: Nature's Electric Generators

Several species of fish, such as electric eels and electric rays, have specialized organs that generate electricity. These organs contain electrocytes, which work similarly to batteries. Electric eels, for example, can produce a shock of up to 600 volts, used for hunting and self-defense.

Interesting Fact: The study of electric fish contributed to the development of early electrical technology. Researchers like Alessandro Volta drew inspiration from these natural generators when inventing the voltaic pile, the precursor to the modern battery.

The Human Body and Electricity

The human body relies on electrical signals for various functions. Nerve cells, or neurons, transmit information through electrical impulses, allowing us to think, move, and sense the world around us. The heart's rhythm is controlled by electrical signals, which can be measured by an electrocardiogram (ECG).

Medical Applications: Electrotherapy, which uses electrical impulses to stimulate nerves and muscles, is a therapeutic technique for pain relief and rehabilitation. Pacemakers, devices that help regulate heartbeats, are another crucial medical application of electricity in the human body.

Lightning: Nature's Spectacular Discharge

Lightning is a dramatic natural display of electricity, occurring during thunderstorms. It results from the buildup of static electricity within clouds. When the electric potential becomes too great, a discharge occurs, creating a bolt of lightning. The accompanying thunder is the sound of rapidly expanding air heated by the lightning's intense energy.

Curiosity: A single lightning bolt can carry a current of up to 200,000 amperes and reach temperatures of around 30,000 Kelvin, much hotter than the surface of the sun.

Wireless Power Transmission

The concept of wireless power transmission has been a topic of research since the early 20th century. Nikola Tesla pioneered experiments in transmitting electricity without wires using resonant inductive coupling. Today, wireless charging technology for devices like smartphones and electric toothbrushes is a practical application of this principle.

Future Prospects: Advancements in wireless power technology could lead to more efficient ways to power electric vehicles and even potentially transmit electricity over longer distances, reducing the need for extensive power grids.

The Electric Age: Transforming Society

The widespread adoption of electricity in the late 19th and early 20th centuries revolutionized society. Electric lighting extended the day, allowing for longer working hours and improving safety. The development of electrical appliances transformed household chores, while the advent of electric transportation systems facilitated urban growth.

Remarkable Transformation: The electric age not only enhanced productivity and convenience but also paved the way for modern communication technologies, from the telegraph to the internet, connecting the world in unprecedented ways.

Renewable Energy and Electricity

As the world seeks sustainable energy solutions, renewable sources of electricity are gaining prominence. Solar panels, wind turbines, and hydroelectric power plants harness natural forces to generate electricity without depleting resources or emitting greenhouse gases.

Insight: The integration of renewable energy into the electrical grid requires advanced technologies for energy storage and distribution. Innovations in battery technology and smart grids are critical for maximizing the potential of renewable energy sources.

The Future of Electricity

The future of electricity is poised to be shaped by ongoing technological advancements and a growing emphasis on sustainability. Smart grids, capable of efficiently managing energy supply and demand, will enhance the reliability and efficiency of power distribution. Emerging technologies like quantum computing and superconductors could revolutionize how we generate, store, and use electricity.

Vision: Imagine a world where clean, renewable energy powers homes, industries, and transportation, reducing our environmental footprint and fostering a sustainable future for generations to come.

References and Further Reading for Chapter 2
Books

"Electricity and Magnetism" by Edward M. Purcell - A comprehensive introduction to the principles of electricity and magnetism.

"Fundamentals of Physics" by David Halliday, Robert Resnick, and Jearl Walker - Provides a solid foundation in the basic concepts of physics, including electricity.

Articles and Publications

"Electric Fields and Forces" (Physics Today) - An article exploring the concepts of electric fields and forces in various contexts.

"Understanding Electric Potential and Voltage" (Scientific American) - An in-depth look at the principles of electric potential and voltage.

Online Resources

Khan Academy - Introduction to Electricity - Video lessons and interactive exercises covering the basics of electricity.

HyperPhysics - Electric Charge and Electric Field - An online resource providing detailed explanations and diagrams of electric charge and fields.

Journals and Periodicals

American Journal of Physics - Articles and papers on various topics in physics, including fundamental electrical principles.

IEEE Transactions on Education - Publications focusing on educational methods and resources in electrical engineering and physics.

Chapter 3

Simple Electrical Circuits

3.1. Basic Components: Resistors, Capacitors, Inductors

Understanding the fundamental components of electrical circuits is crucial for anyone beginning their journey in electrical engineering. This section will cover resistors, capacitors, and inductors, which are the building blocks of more complex circuits. At the end of the book, you can find a complete table with the symbols for the most common electrical and electronic components.

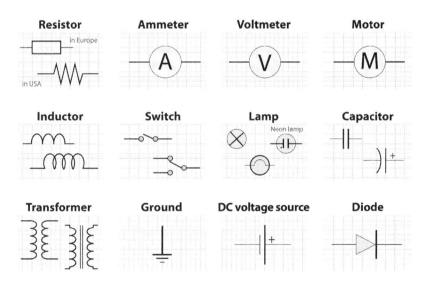

Simple Electrical Circuits

Resistors

Resistors are passive electrical components that oppose the flow of electric current. Their primary function is to limit the amount of current that can pass through a circuit. The symbol for a resistor in a circuit diagram is a zigzag line (USA) or a rectangular box (International).

Types of Resistors

Fixed Resistors: These have a set resistance value that does not change. Common types include carbon film, metal film, and wire-wound resistors.

Variable Resistors: Also known as potentiometers or rheostats, these allow for adjustment of resistance. They are used in applications like volume control in audio devices.

Ohm's Law and Resistors

As discussed in Chapter 2, Ohm's Law is fundamental in understanding how resistors operate in circuits. According to Ohm's Law:

$$V = IR$$

where V is the voltage across the resistor, I is the current through the resistor, and R is the resistance.

Applications of Resistors

Resistors are used in various applications, such as:

- **Current Limiting**: Protecting components from excessive current.
- **Voltage Dividers**: Creating specific voltage levels.
- **Pull-Up/Pull-Down Resistors**: Ensuring known states in digital circuits.

Capacitors

Capacitors are passive components that store and release electrical energy. They consist of two conductive plates separated by an insulating material called a dielectric. The symbol for a capacitor in circuit diagrams is two parallel lines (one curved if it's a polarized capacitor).

Types of Capacitors

Ceramic Capacitors: Made from ceramic materials, these capacitors are widely used due to their small size and stability.

Electrolytic Capacitors: These have a larger capacitance value and are polarized, meaning they have a positive and negative terminal.

Tantalum Capacitors: Known for their reliability and long life, these are often used in sensitive electronic equipment.

Capacitance and Charge Storage

The capacitance (C) of a capacitor is measured in farads (F) and indicates the amount of charge it can store per unit voltage:

$$C = \frac{Q}{V}$$

where Q is the charge in coulombs and V is the voltage.

Applications of Capacitors

Capacitors are used in numerous applications, such as:

- **Energy Storage**: Providing power in applications like camera flashes.
- **Filtering**: Smoothing out voltage fluctuations in power supplies.
- **Tuning Circuits**: Used in radio frequency applications to select specific frequencies.

Inductors

Inductors are passive components that store energy in a magnetic field when electric current flows through them. They consist of a coil of wire and are represented by a coiled symbol in circuit diagrams.

Types of Inductors

Air Core Inductors: These have no core material and are used in high-frequency applications.

Iron Core Inductors: These have an iron core to increase inductance and are used in low-frequency applications.

Ferrite Core Inductors: These use ferrite material as the core and are common in high-frequency applications due to their low losses.

Inductance and Magnetic Fields

The inductance (L) of an inductor is measured in henries (H) and relates to the ability of the inductor to store energy in a magnetic field:

$$V = L \frac{dI}{dt}$$

where V is the induced voltage, L is the inductance, and $\frac{dI}{dt}$ is the rate of change of current.

Applications of Inductors

Inductors are used in various applications, such as:

- **Energy Storage**: In power supplies and converters.
- **Filtering**: In conjunction with capacitors to filter out unwanted frequencies.
- **Transformers**: Inductors are key components in transformers, which transfer energy between circuits through electromagnetic induction.

Combining Components

In practical circuits, resistors, capacitors, and inductors are often combined to achieve desired electrical characteristics. For example:

- **RC Circuits**: Combinations of resistors and capacitors are used in timing and filtering applications.

- **RL Circuits**: Combinations of resistors and inductors are used in filtering and tuning applications.

- **RLC Circuits**: Circuits that include resistors, capacitors, and inductors are used in more complex filtering and resonance applications.

Understanding how these components interact is crucial for designing effective circuits. For detailed examples and practical applications, see the upcoming sections on series and parallel circuits.

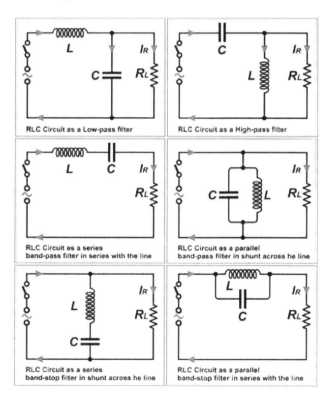

Simple Electrical Circuits

3.2. Series and Parallel Circuits

Understanding how components like resistors, capacitors, and inductors behave when connected in series and parallel is fundamental to analyzing and designing electrical circuits. This section explores the characteristics and calculations involved in these configurations.

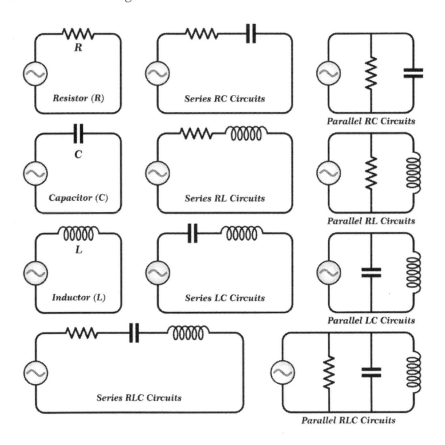

Series Circuits

Definition and Characteristics

In a series circuit, components are connected end-to-end in such a way that there is only one path for current to flow. This means that the same current passes through each component in the circuit.

Key Points:

- **Current**: The current (I) is the same through all components in a series circuit.

- **Voltage**: The total voltage (V_{total}) across a series circuit is the sum of the voltages across each component: $V_{total} = V_1 + V_2 + V_3 + ...$

- **Resistance**: The total resistance (R_{total}) in a series circuit is the sum of the resistances of each component: $R_{total} = R_1 + R_2 + R_3 + ...$

Example Calculation: Series Resistors

Consider a series circuit with three resistors:

- $R_1 = 2k\Omega$
- $R_2 = 3k\Omega$
- $R_3 = 5k\Omega$

The total resistance would be: $R_{total} = 2k\Omega + 3k\Omega + 5k\Omega = 10k\Omega$

If the circuit is connected to a 10V power supply, the current through the circuit would be:

$$I = \frac{V_{total}}{R_{total}} = \frac{10V}{10k\Omega} = 1mA$$

Simple Electrical Circuits

Voltage Drop in Series Circuits

In a series circuit, the voltage drop across each resistor can be calculated using Ohm's Law:

$$V_n = I \times R_n$$

For the resistors in the example above:

- $V_1 = 1\text{mA} \times 2\text{k}\Omega = 2\text{V}$
- $V_2 = 1\text{mA} \times 3\text{k}\Omega = 3\text{V}$
- $V_3 = 1\text{mA} \times 5\text{k}\Omega = 5\text{V}$

The total of these voltage drops equals the total voltage of the power supply.

Parallel Circuits

Definition and Characteristics

In a parallel circuit, components are connected across the same two points, creating multiple paths for current to flow. This means that the voltage across each component is the same, but the current can vary.

Key Points:

- **Voltage**: The voltage (V) is the same across all components in a parallel circuit.

- **Current**: The total current (I_{total}) is the sum of the currents through each parallel branch: $I_{total} = I_1 + I_2 + I_3 + ...$

- **Resistance**: The total resistance (R_{total}) in a parallel circuit is less than the smallest resistance and is calculated using the reciprocal formula:
$$\frac{1}{R_{total}} = \frac{1}{R_1} + \frac{1}{R_2} + \frac{1}{R_3} + ...$$

Example Calculation: Parallel Resistors

Consider a parallel circuit with three resistors:

- $R_1 = 2\text{k}\Omega$
- $R_2 = 3\text{k}\Omega$
- $R_3 = 6\text{k}\Omega$

The total resistance would be:

$$\frac{1}{R_{total}} = \frac{1}{2\text{k}\Omega} + \frac{1}{3\text{k}\Omega} + \frac{1}{6\text{k}\Omega} = \frac{3+2+1}{6\text{k}\Omega} = \frac{6}{6\text{k}\Omega} = \frac{1}{1\text{k}\Omega}$$

$$R_{total} = 1\text{k}\Omega$$

If the circuit is connected to a 10V power supply, the total current would be:

$$I_{total} = \frac{V}{R_{total}} = \frac{10\text{V}}{1\text{k}\Omega} = 10\text{mA}$$

Current in Parallel Circuits

The current through each resistor in a parallel circuit can be calculated using Ohm's Law:

$$I_n = \frac{V}{R_n}$$

For the resistors in the example above:

- $I_1 = \frac{10\text{V}}{2\text{k}\Omega} = 5\text{mA}$
- $I_2 = \frac{10\text{V}}{3\text{k}\Omega} = 3.33\text{mA}$
- $I_3 = \frac{10\text{V}}{6\text{k}\Omega} = 1.67\text{mA}$

The sum of these currents equals the total current in the circuit:

$$I_{total} = 5\text{mA} + 3.33\text{mA} + 1.67\text{mA} = 10\text{mA}$$

Simple Electrical Circuits

Combining Series and Parallel Circuits

In more complex circuits, components can be combined in both series and parallel configurations. Understanding how to calculate the total resistance, voltage, and current in these mixed circuits is essential for designing functional electrical systems.

Example: Mixed Circuit

Consider a circuit where two resistors (R_1 and R_2) are connected in series, and this combination is connected in parallel with another resistor (R_3):

Calculate the Series Resistance:

$$R_{12} = R_1 + R_2$$

Calculate the Total Resistance:

$$\frac{1}{R_{total}} = \frac{1}{R_{12}} + \frac{1}{R_3}$$

Determine the Total Current using the total resistance and applied voltage.

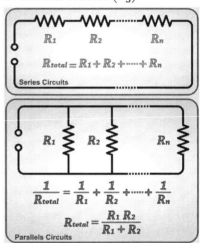

Practical Applications

Series and parallel circuits are used in countless applications:

- **Series Circuits**: Often used in string lights where the failure of one bulb affects the entire string.

- **Parallel Circuits**: Common in household wiring, where each outlet receives the same voltage, allowing multiple devices to operate independently.

Understanding these configurations enables engineers and technicians to design circuits that meet specific needs, whether for simple devices or complex electronic systems.

3.3. Ohm's Law and Kirchhoff's Laws

Ohm's Law and Kirchhoff's Laws are foundational principles in electrical engineering, crucial for analyzing and understanding the behavior of electrical circuits. This section will explore these laws in detail and demonstrate how they are applied to both simple and complex circuits.

Ohm's Law

Definition and Formula

Ohm's Law, named after the German physicist Georg Simon Ohm, describes the relationship between voltage (V), current (I), and resistance (R) in a circuit. The law is stated as:

$$V = I \times R$$

This equation implies that the voltage across a resistor is directly proportional to the current flowing through it and the resistance of the resistor.

Applications of Ohm's Law

Ohm's Law is widely used to:

- **Calculate Voltage**: If the current and resistance are known, the voltage across a component can be determined.

- **Calculate Current**: If the voltage and resistance are known, the current flowing through a circuit can be found.

- **Calculate Resistance**: If the voltage and current are known, the resistance of a component can be calculated.

Example Calculation Using Ohm's Law

Consider a simple circuit with a 9V battery and a resistor of 3Ω:

- The current (I) flowing through the circuit would be:

$$I = \frac{V}{R} = \frac{9V}{3\Omega} = 3A$$

This calculation shows how Ohm's Law can be used to determine the current in a circuit when the voltage and resistance are known.

Simple Electrical Circuits

Limitations of Ohm's Law

While Ohm's Law is fundamental, it applies primarily to ohmic materials—those that have a constant resistance regardless of the applied voltage. Non-ohmic materials, such as diodes and transistors, do not follow Ohm's Law directly because their resistance varies with voltage and current.

Kirchhoff's Laws

Gustav Kirchhoff, a German physicist, formulated two essential laws that extend the principles of Ohm's Law to more complex circuits: Kirchhoff's Current Law (KCL) and Kirchhoff's Voltage Law (KVL).

Kirchhoff's Current Law (KCL)

Definition

Kirchhoff's Current Law states that the total current entering a junction (or node) in an electrical circuit must equal the total current leaving the junction. Mathematically, this is expressed as:

$$\sum I_{in} = \sum I_{out}$$

This law is based on the principle of conservation of charge, ensuring that all charge entering a junction must leave it.

Application of KCL

KCL is used to analyze circuits with multiple branches and nodes, helping to determine the current distribution in complex networks.

Example: Applying KCL

Consider a node in a circuit where three currents meet:

- $I_1 = 5A$ (entering)
- $I_2 = 2A$ (entering)
- $I_3 = ?$ (leaving)

According to KCL:

$$I_3 = I_1 + I_2 = 5A + 2A = 7A$$

This shows that the current leaving the node must equal the sum of the currents entering it.

Kirchhoff's Voltage Law (KVL)

Definition

Kirchhoff's Voltage Law states that the sum of the electrical potential differences (voltage) around any closed loop or mesh in a circuit must equal zero. Mathematically, this is expressed as:

$$\sum V = 0$$

This law is based on the principle of conservation of energy, which ensures that the total energy gained by charges in a loop must equal the total energy lost.

Application of KVL

KVL is used to analyze the voltage distribution in circuits with multiple loops, helping to determine the voltage drops across various components.

Example: Applying KVL

Consider a simple loop with three components:

- A battery with $V_1 = 12$ V
- A resistor with a voltage drop of $V_2 = 7$ V
- Another resistor with a voltage drop of $V_3 = ?$

According to KVL:

$$V_1 - V_2 - V_3 = 0$$
$$V_3 = V_1 - V_2 = 12V - 7V = 5V$$

This shows that the sum of voltage drops in the loop equals the voltage provided by the battery.

Simple Electrical Circuits

Combining Ohm's Law with Kirchhoff's Laws

In circuit analysis, Ohm's Law is often combined with Kirchhoff's Laws to solve for unknown values in complex circuits. By applying KCL at junctions and KVL in loops, along with Ohm's Law for individual components, engineers can systematically solve for currents, voltages, and resistances in intricate electrical networks.

Example: Analyzing a Complex Circuit

Consider the circuit shown in the image, which features two loops, a light bulb (represented by a 4Ω resistor), and multiple resistors connected to two different voltage sources.

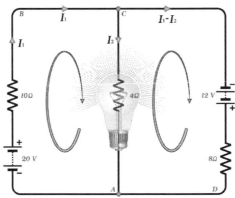

To analyze this circuit, follow these systematic steps:

Apply Kirchhoff's Current Law (KCL) at Junction C:

At junction C, where the currents I_1 and I_2 split, apply KCL. According to KCL:

$$I_1 = I_2 + I_3$$

$$I_3 = I_1 - I_2$$

$$I_1 = I_2 + (I_1 - I_2)$$

This equation ensures that the total current entering junction C equals the total current leaving it.

Apply Kirchhoff's Voltage Law (KVL) to Each Loop:

Left Loop (ABC):

Starting at point A and moving clockwise through the loop ABC:

$$20V - (10\Omega \times I_1) - (4\Omega \times I_2) = 0$$

This equation helps determine the relationship between the current I_1 through the 10Ω resistor and the current I_2 through the 4Ω resistor (light bulb) in the left loop.

Right Loop (ACD):

Starting at point A and moving clockwise through the loop ACD:

$$12V - (8\Omega \times (I_1 - I_2)) + (4\Omega \times I_2) = 0$$

In this equation, the term $(4\Omega \times I_2)$ is added instead of subtracted, reflecting that I_2 flows in the opposite direction through the 4Ω resistor in this loop.

Use Ohm's Law to Solve for Unknown Values:

With the currents I_1 and I_2 determined from KCL and KVL, use Ohm's Law $V = IR$ to calculate the voltage drops across individual resistors and verify the calculated currents.

By applying KCL at junction C and KVL around both loops ABC and ACD, and then solving the resulting equations with Ohm's Law, you can systematically determine the unknown currents and voltages within this complex circuit. This approach is essential for understanding and analyzing intricate electrical networks.

Practical Applications of Kirchhoff's Laws

Kirchhoff's Laws are indispensable in designing and analyzing:

- **Power Distribution Networks**: Ensuring that currents and voltages are correctly balanced in power systems.
- **Electronic Circuits**: Used in the design of amplifiers, filters, and other electronic devices.
- **Safety Systems**: Ensuring that all parts of a circuit operate within safe limits by accurately calculating voltage drops and current flows.

3.4. Quiz for Review

Access Additional Quiz Material

Scan the QR code below to download a PDF file with additional quiz questions. This allows you to practice offline and further reinforce your understanding of the concepts covered in the book.

3.5. Practical Exercises

Download Additional Exercises

Scan the QR code below to download a PDF file containing additional practical exercises. These exercises will help you apply the concepts learned in this book and further enhance your hands-on skills.

3.6. Curiosities and Further Insights

As we delve deeper into the world of electrical circuits, there are fascinating aspects and real-world applications that extend beyond the basics. This section explores interesting facts, historical anecdotes, and advanced insights related to simple electrical circuits, offering a broader perspective on how these concepts shape our everyday lives.

The Origins of Circuit Diagrams

Circuit diagrams, also known as schematic diagrams, are the universal language used by engineers and technicians worldwide to design, document, and communicate the construction of electrical systems. The use of standardized symbols began in the early 20th century, simplifying the process of understanding and building circuits.

Historical Note: The development of circuit diagrams is closely linked to the rise of electrical engineering as a profession. Early pioneers like Thomas Edison and Nikola Tesla relied on hand-drawn schematics to design some of the first electrical power systems. These diagrams were crucial in the development of technologies that powered the modern world.

Household Wiring: A Hidden Marvel

The wiring behind the walls of our homes is a hidden but essential system that makes modern life possible. Household wiring follows strict codes and standards to ensure safety and efficiency. Most residential wiring uses a combination of series and parallel circuits to distribute power to various outlets, lights, and appliances.

Curiosity: Have you ever wondered why the lights in your home remain on when one bulb burns out? This is because they are connected in parallel, ensuring that the circuit remains complete even if one component fails. This design increases reliability and convenience, making it a critical aspect of modern electrical systems.

The Paradox of Christmas Lights

A common holiday frustration is dealing with a string of Christmas lights where one burnt-out bulb causes the entire string to fail. This occurs in traditional series-wired lights, where the failure of one bulb breaks the circuit. Newer designs often incorporate parallel wiring or shunt paths to prevent this issue, demonstrating the practical application of parallel circuits to enhance reliability.

Electrical Safety: The Importance of Grounding

Grounding is a critical safety measure in electrical systems, providing a safe path for electricity to flow back to the ground in the event of a fault, preventing electric shocks and reducing the risk of fire. Ground Fault Circuit Interrupters (GFCIs) are another safety innovation that quickly cuts off power if an imbalance between the hot and neutral wires is detected.

Insight: Grounding and GFCIs are essential in areas where electricity and water may come into contact, such as kitchens, bathrooms, and outdoor outlets. These safety features are now required by electrical codes in many countries, reflecting the evolution of safety standards in electrical engineering.

The Practicality of Voltage Dividers

Voltage dividers are simple yet powerful tools in electrical engineering, used to reduce the voltage to a desired level. A practical application of a voltage divider is in adjusting the input voltage for sensors and analog-to-digital converters (ADCs) in electronic devices.

Application Example: In a simple temperature sensor circuit, a thermistor might be connected in series with a fixed resistor to form a voltage divider. The output voltage from the divider is then proportional to the temperature, allowing for accurate readings by the ADC. This demonstrates how voltage dividers are integral to sensor applications.

The Efficiency of Modern Home Wiring

Modern home electrical systems are designed with efficiency and safety in mind, utilizing a combination of series and parallel circuits to optimize performance. Circuit breakers, ground fault circuit interrupters (GFCIs), and properly rated wiring materials ensure that homes are both energy-efficient and safe.

Green Technologies: As homes become more energy-conscious, innovations such as smart lighting systems, energy-efficient appliances, and solar panel installations are becoming more common. These systems often rely on sophisticated circuit designs that incorporate both series and parallel elements to maximize efficiency and reduce energy consumption.

The Future of Home Circuits: Smart Homes and IoT

As technology advances, the concept of smart homes is becoming a reality. Smart home systems integrate various devices and appliances, allowing them to communicate with each other and be controlled remotely. These systems often rely on both wired and wireless circuits, combining traditional electrical engineering with cutting-edge technology.

Vision: Imagine controlling your home's lighting, heating, and security systems from your smartphone, even when you're not at home. Smart home technology, powered by the Internet of Things (IoT), is making this possible, offering convenience, energy savings, and enhanced security.

The Role of Circuit Breakers and Fuses

Circuit breakers and fuses are crucial for protecting electrical circuits from damage caused by overcurrent or short circuits. These devices interrupt the flow of electricity when they detect an overload, preventing potential hazards like fires.

Curiosity: Did you know that circuit breakers can be reset after they trip, but fuses must be replaced once they blow? Circuit breakers are commonly used in modern homes because they are more convenient and offer better protection.

Learning from Failures: The Importance of Troubleshooting

In electrical engineering, troubleshooting is an essential skill. Even the best-designed circuits can encounter issues, and understanding how to diagnose and fix these problems is key to maintaining reliable systems. Troubleshooting involves systematically testing different parts of a circuit to identify the cause of a malfunction.

Historical Insight: Thomas Edison, who held over 1,000 patents, often emphasized the importance of learning from failures. He famously tested thousands of materials before finding the right filament for the incandescent light bulb, demonstrating the value of persistence and problem-solving in engineering.

The Discovery of Ohm's Law

Georg Simon Ohm, a German physicist, formulated Ohm's Law in 1827. His work was initially met with skepticism by the scientific community, as the idea of a simple linear relationship between voltage, current, and resistance was revolutionary at the time. Despite the initial resistance, Ohm's Law became one of the fundamental principles of electrical engineering.

Impact on Circuit Design: Ohm's Law provides a straightforward method for calculating the behavior of electrical circuits, making it an indispensable tool for engineers. Whether designing a simple circuit or a complex electrical system, Ohm's Law serves as the foundation for understanding how different components interact.

References and Further Reading for Chapter 3

Books

"Practical Electronics for Inventors" by Paul Scherz and Simon Monk - *A comprehensive guide that includes practical circuit design tips and in-depth explanations of components.*

Articles and Publications

"The History of Ohm's Law" (Physics Today) - *An article exploring the development and significance of Ohm's Law in electrical engineering.*

"Voltage Dividers in Modern Electronics" (IEEE Spectrum) - *A detailed look at how voltage dividers are used in various electronic applications.*

Online Resources

Khan Academy - Circuit Basics - *Video tutorials and exercises covering the basics of series and parallel circuits.*

HyperPhysics - Ohm's Law and Circuits - *An online resource with diagrams and explanations for understanding Ohm's Law and circuit analysis.*

Journals and Periodicals

Journal of Electrical Engineering and Technology - *Research articles and case studies on modern circuit design and electrical systems.*

IEEE Transactions on Circuits and Systems - *Papers focused on the latest advancements in circuit theory and applications.*

Chapter 4

Magnetism and Electromagnetism

Introduction to Magnetism

Magnetism is a fundamental force of nature that plays a crucial role in many aspects of electrical engineering. Understanding the principles of magnetism is essential for comprehending how electric currents interact with magnetic fields, leading to the development of various technologies, from simple compasses to advanced electrical motors and generators.

What is Magnetism?

Magnetism is the force exerted by magnets when they attract or repel each other. This force is caused by the movement of electric charges, specifically the movement of electrons within atoms. At the most basic level, magnetism arises from the spin and orbital motion of electrons, which creates a magnetic dipole moment. In materials where these magnetic moments align in a consistent direction, the material exhibits magnetic properties.

Magnetic Poles

Every magnet has two poles: a north pole and a south pole. The fundamental law of magnetism states that like poles repel each other, while unlike poles attract each other. This behavior is similar to electric charges, where like charges repel and opposite charges attract—a dualism you may recall from Chapter 2: Fundamentals of Electricity.

When a magnet is freely suspended, such as in a compass, its north pole will naturally align with the Earth's magnetic north. This principle is used in navigation, where a magnetic compass helps determine direction.

Magnetic Fields

A magnetic field is the region around a magnet where magnetic forces can be detected. The strength and direction of a magnetic field are represented by magnetic field lines, which flow from the north pole to the south pole of a magnet. The density of these lines indicates the strength of the magnetic field: the closer the lines, the stronger the magnetic field.

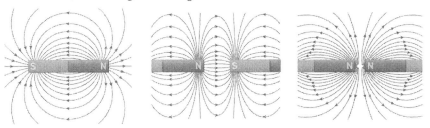

Mathematically, the magnetic field is denoted by the symbol B and is measured in teslas (T). The concept of a magnetic field is analogous to the electric field discussed in Chapter 2, where electric charges create electric fields that influence other charges in the vicinity.

Magnetic Materials

Materials respond differently to magnetic fields based on their atomic structure. There are three main categories of magnetic materials:

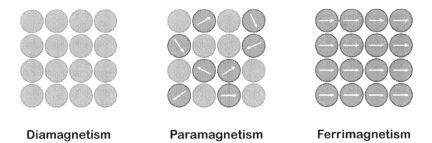

Diamagnetism   **Paramagnetism**   **Ferrimagnetism**

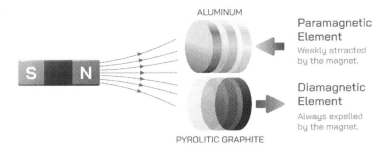

Ferromagnetic Materials: These materials, such as iron, cobalt, and nickel, exhibit strong magnetic properties. In ferromagnetic materials, the magnetic moments of atoms align in parallel, leading to a net magnetic field. These materials are commonly used to make permanent magnets.

Paramagnetic Materials: These materials, like aluminum and platinum, have a weaker response to magnetic fields. The magnetic moments of atoms in paramagnetic materials do not align as strongly, resulting in a weaker overall magnetic field.

Diamagnetic Materials: Diamagnetic materials, such as copper and gold, are repelled by magnetic fields. The magnetic moments in diamagnetic materials tend to align opposite to the applied magnetic field, causing a weak repulsive effect.

Understanding the properties of these materials is essential for designing and applying magnetic technologies, particularly in creating electromagnets, which will be explored further in the subsequent sections of this chapter.

The Earth's Magnetic Field

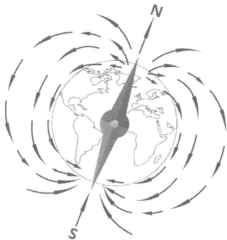

The Earth itself acts as a giant magnet, with a magnetic field that extends from its interior out into space. This geomagnetic field is generated by the movement of molten iron within the Earth's outer core, which creates electric currents and, consequently, a magnetic field. The Earth's magnetic field is what causes compass needles to point north, aiding navigation for centuries.

Connection to Electricity

Magnetism and electricity are closely related, as evidenced by the fact that a moving electric charge generates a magnetic field. This relationship is at the heart of electromagnetism, which describes how electric currents and magnetic fields interact. The discovery of this relationship led to the development of numerous technologies, including electric motors, generators, and transformers—all of which will be discussed in greater detail in later sections.

For example, the principle that a current-carrying conductor generates a magnetic field is the basis for electromagnets, which are vital components in many electrical devices. This connection between electricity and magnetism will be further explored in the next section of this chapter, where we delve into how magnetic fields are generated by electric currents.

4.1. Magnetic Fields Generated by Current

The relationship between electricity and magnetism is one of the cornerstones of modern physics and electrical engineering. When an electric current flows through a conductor, it generates a magnetic field around it. This phenomenon is the basis for electromagnetism This phenomenon is the basis for electromagnetism and plays a crucial role in various applications.

The Discovery of Electromagnetism

The connection between electricity and magnetism was first discovered by Danish physicist Hans Christian Ørsted in 1820.

Ørsted observed that a compass needle was deflected when placed near a current-carrying wire, indicating that the electric current was generating a magnetic field. This discovery was groundbreaking because it demonstrated that electricity and magnetism are interrelated phenomena, which had previously been considered separate.

Magnetic Field Around a Straight Conductor

When a current flows through a straight conductor, such as a wire, it generates a magnetic field around the conductor. The direction of this magnetic field can be determined using the **right-hand rule**: if you point the thumb of your right hand in the direction of the current, your fingers will curl in the direction of the magnetic field lines.

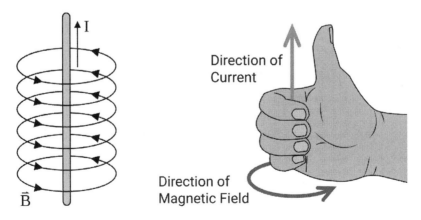

Characteristics of the Magnetic Field:

- The magnetic field forms concentric circles around the conductor.
- The strength of the magnetic field decreases with distance from the conductor.
- The magnetic field is stronger when the current is higher.

Mathematically, the magnetic field B around a long, straight conductor can be calculated using Ampère's Law:

$$B = \frac{\mu_0 I}{2\pi r}$$

where:

- B is the magnetic field in teslas (T),
- μ_0 is the permeability of free space ($4\pi \times 10^{-7}$ T·m/A)
- I is the current in amperes (A),
- r is the distance from the conductor in meters (m).

Magnetic Field Around a Circular Loop

If a current-carrying conductor is bent into a circular loop, the magnetic field produced by each segment of the loop adds together, creating a stronger magnetic field in the center of the loop. The direction of the magnetic field inside the loop can also be determined using the right-hand rule: if your fingers curl in the direction of the current, your thumb will point in the direction of the magnetic field.

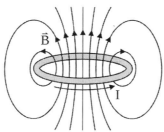

The magnetic field at the center of a circular loop is given by:

$$B = \frac{\mu_0 I}{2R}$$

where:
- B is the magnetic field at the center of the loop,
- R is the radius of the loop.

Solenoids and Electromagnets

A solenoid is a coil of wire, typically wound into a cylindrical shape, that generates a magnetic field when an electric current passes through it. The magnetic field produced by a solenoid is similar to that of a bar magnet, with a clear north and south pole. The strength of the magnetic field inside a solenoid is directly proportional to the number of turns in the coil and the current passing through it.

The magnetic field inside a long solenoid is uniform and parallel to the axis of the solenoid and is given by:

$$B = \mu_0 n I$$

where:
- n represents the number of turns per unit length, specifically the number of turns per meter along the length of the solenoid.

When a solenoid is used in combination with a ferromagnetic core (such as iron), it becomes an electromagnet. The ferromagnetic core significantly increases the strength of the magnetic field, making electromagnets extremely powerful compared to simple solenoids. Electromagnets are widely used in various applications, including electric motors, relays, and magnetic cranes used in scrapyards to lift heavy metal objects.

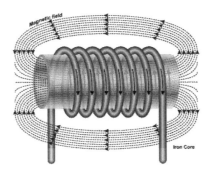

4.2. Electromagnets and Their Applications

Electromagnets are one of the most versatile and widely used applications of electromagnetism. Unlike permanent magnets, which maintain a constant magnetic field, electromagnets generate a magnetic field only when an electric current flows through them. This controllable nature of electromagnets makes them essential in a wide range of applications, from industrial machinery to everyday electronic devices.

What is an Electromagnet?

An electromagnet consists of a coil of wire, usually wound around a ferromagnetic core, such as iron (see previous image). When an electric current passes through the coil, it generates a magnetic field. The strength of this magnetic field can be adjusted by varying the amount of current or the number of turns in the coil. When the current is turned off, the magnetic field disappears, making electromagnets highly adaptable for various tasks.

How Electromagnets Work

The operation of an electromagnet is based on the principles of electromagnetism that we explored in the previous section. The electric current flowing through the coil creates a magnetic field around the wire. This field is concentrated and amplified by the ferromagnetic core, which becomes magnetized and significantly increases the overall strength of the magnetic field.

The direction of the magnetic field in an electromagnet can be determined by the right-hand rule, as discussed earlier. The poles of the electromagnet can also be reversed by changing the direction of the current flow, providing further control over its operation.

Key Factors Affecting Electromagnet Strength:

- **Number of Coil Turns**: More turns in the coil increase the magnetic field strength.

- **Current Strength**: Higher current through the coil results in a stronger magnetic field.

- **Core Material:** The use of a ferromagnetic core, such as iron, amplifies the magnetic field, making the electromagnet much more powerful.

Applications of Electromagnets

Electromagnets are used in countless applications across various industries. Below are some of the most common and significant uses of electromagnets.

Electric Motors

Electric motors, which convert electrical energy into mechanical energy, rely on electromagnets to create the rotational force needed to turn the motor's shaft. In a typical motor, the electromagnet is positioned within a magnetic field generated by either permanent magnets or other electromagnets. When current flows through the electromagnet, it generates a magnetic field that interacts with the surrounding magnetic field, causing the motor's rotor to turn.

Electric motors are ubiquitous, found in everything from household appliances like washing machines and fans to industrial equipment and electric vehicles.

Transformers

Transformers, which are essential for transmitting electricity over long distances, use electromagnets to transfer electrical energy between circuits. In a transformer, an alternating current (AC) passes through the primary coil, generating a magnetic field that induces a current in the secondary coil. This process allows for the efficient transfer of energy and the adjustment of voltage levels to suit different needs.

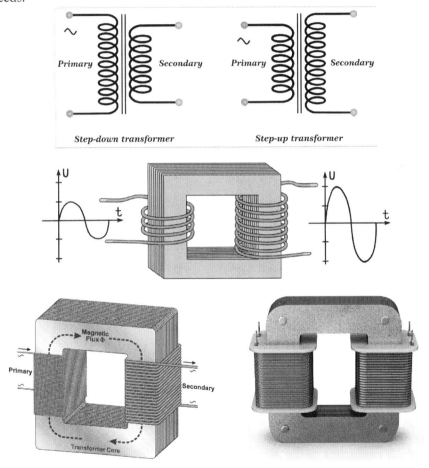

Transformers are critical in the power distribution network, enabling the delivery of electricity from power plants to homes and businesses.

Relays and Switches

Electromagnetic relays are switches that use electromagnets to control the operation of circuits. When current flows through the relay's coil, it generates a magnetic field that moves a metal armature, opening or closing the circuit's contacts. Relays are used in a variety of applications, including automotive systems, industrial controls, and telecommunications.

The ability of relays to control high-power circuits with low-power signals makes them indispensable in modern electronics.

Magnetic Lifting Devices

Electromagnets are commonly used in industrial settings for lifting and moving heavy metal objects, such as scrap metal and steel beams. These magnetic lifting devices can be turned on and off as needed, providing a safe and efficient method for handling large, heavy materials.

In scrapyards, for example, electromagnets are used to sort and transport metallic materials, making the recycling process more efficient.

Magnetic Resonance Imaging (MRI) Machines

In the medical field, electromagnets play a crucial role in Magnetic Resonance Imaging (MRI) machines. These devices use powerful electromagnets to generate a magnetic field that aligns the protons in the body's hydrogen atoms.

When exposed to radio waves, these protons emit signals that are used to create detailed images of the body's internal structures.
MRI machines have revolutionized medical diagnostics, allowing for non-invasive imaging of soft tissues and organs.

Electromagnetic Brakes

Electromagnetic brakes are used in various applications, from trains to industrial machinery. In these systems, an electromagnet generates a magnetic field that creates friction, slowing down or stopping the movement of a rotating shaft. These brakes are highly reliable and can be controlled with precision, making them ideal for use in safety-critical applications.

Audio Speakers

The speakers in audio systems use electromagnets to convert electrical signals into sound. In a speaker, an electromagnet (called the voice coil) is attached to a cone. When an audio signal passes through the coil, it generates a magnetic field that interacts with the magnetic field of a permanent magnet, causing the coil and the attached cone to move. This movement pushes air, creating sound waves that we hear as music or speech.

Electromagnets vs. Permanent Magnets

While both electromagnets and permanent magnets generate magnetic fields, there are significant differences between them. Permanent magnets have a constant magnetic field and do not require a power source, making them suitable for applications where a steady magnetic field is needed. However, they lack the flexibility and control offered by electromagnets.

Electromagnets, on the other hand, can be turned on and off, and their strength can be adjusted as needed. This makes them ideal for applications where a variable or controlled magnetic field is required.

Electromagnets are integral to modern technology, enabling the operation of countless devices and systems that we rely on every day. Their ability to generate a controlled magnetic field with adjustable strength makes them indispensable in fields ranging from industrial manufacturing to medical imaging. Understanding how electromagnets work and their diverse applications provides valuable insight into the role of electromagnetism in our daily lives.

4.3. Quiz for Review

Access Additional Quiz Material

Scan the QR code below to download a PDF file with additional quiz questions. This allows you to practice offline and further reinforce your understanding of the concepts covered in the book.

4.4. Practical Exercises

Download Additional Exercises

Scan the QR code below to download a PDF file containing additional practical exercises. These exercises will help you apply the concepts learned in this book and further enhance your hands-on skills.

4.5. Curiosities and Further Insights

Magnetism and electromagnetism are not just theoretical concepts; they are integral to many fascinating phenomena and technologies that shape our modern world. In this section, we'll explore some intriguing aspects and real-world applications of these forces, as well as some historical anecdotes that highlight their significance.

The Earth as a Giant Magnet

One of the most remarkable examples of magnetism is the Earth itself. The Earth behaves like a giant bar magnet, with a magnetic field that extends from its interior out into space. This geomagnetic field is what causes compass needles to point north, aiding navigation for centuries. The origin of the Earth's magnetic field is believed to be the movement of molten iron in the Earth's outer core, which generates electric currents and, consequently, a magnetic field.

Curiosity: The Earth's magnetic field is not static; it changes over time. This phenomenon, known as geomagnetic reversal, has occurred many times in the planet's history, where the north and south magnetic poles have swapped places. Although these reversals happen over thousands of years, they are a fascinating testament to the dynamic nature of the Earth's magnetism.

The Invention of the Electric Motor

The invention of the electric motor is one of the most significant milestones in the application of electromagnetism. Michael Faraday, an English scientist, is credited with building the first simple electric motor in 1821. Faraday discovered that when a current-carrying wire was placed near a magnet, it experienced a force, causing it to move. This principle is the basis of all electric motors, which convert electrical energy into mechanical motion.

Historical Note: Faraday's experiments laid the foundation for the development of modern electrical engineering. His work demonstrated the practical applications of electromagnetic forces, paving the way for the widespread use of electric motors in countless devices, from household appliances to industrial machinery.

Electromagnetic Waves and Communication

Electromagnetic waves, which are oscillating electric and magnetic fields, are fundamental to modern communication technologies. These waves travel at the speed of light and are used to transmit information over long distances. Radio, television, and mobile phone signals are all examples of electromagnetic waves in action.

Interesting Fact: James Clerk Maxwell, a Scottish physicist, developed the theory of electromagnetism in the 19th century, showing that light itself is an electromagnetic wave. Maxwell's equations describe how electric and magnetic fields interact and propagate through space, forming the basis of much of modern physics and engineering.

Magnetic Levitation and Transportation

Magnetic levitation, or maglev, is an advanced technology that uses powerful electromagnets to lift and propel vehicles without direct contact with the ground. This technology is most famously used in maglev trains, which can travel at incredibly high speeds with minimal friction.

Curiosity: The Shanghai Maglev Train, which operates in China, is one of the fastest commercial trains in the world, reaching speeds of up to 430 km/h (267 mph). The train uses superconducting magnets to achieve levitation and propulsion, making it a marvel of modern engineering.

The Role of Electromagnets in Particle Accelerators

Particle accelerators, such as the Large Hadron Collider (LHC) in Switzerland, use powerful electromagnets to steer and accelerate charged particles to nearly the speed of light. These machines are used to study the fundamental properties of matter and have led to groundbreaking discoveries in physics.

Interesting Fact: The electromagnets used in the LHC are superconducting, meaning they have zero electrical resistance when cooled to extremely low temperatures. This allows them to generate incredibly strong magnetic fields necessary for guiding particles through the accelerator's ring.

The Magnetic North and the Pole Shift

The magnetic north pole is not fixed; it constantly drifts due to changes in the Earth's core. In recent years, the magnetic north has been moving at an accelerated pace, prompting updates to navigation systems that rely on magnetic compasses.

Curiosity: The magnetic north pole has shifted by hundreds of kilometers over the past century. This movement is closely monitored by scientists as it can impact everything from navigation to the accuracy of global positioning systems (GPS).

Electromagnetism in Everyday Life

Electromagnetism is a part of our daily lives in ways we often take for granted. From the speakers in our phones and headphones to the induction cooktops in our kitchens, electromagnets play a crucial role in many modern conveniences.

Application Example: Induction cooktops use electromagnetic induction to heat pots and pans directly, rather than heating the cooktop surface itself. This method is not only efficient but also safer, as the cooktop remains relatively cool to the touch.

References and Further Reading for Chapter 4

Books

"Faraday, Maxwell, and the Electromagnetic Field" by Nancy Forbes and Basil Mahon - A historical account of the discovery and development of electromagnetic theory.

Articles and Publications

"Geomagnetic Reversals" (Nature) - A detailed article on the phenomenon of geomagnetic reversals and their impact on the Earth.

"Magnetic Levitation Trains" (IEEE Spectrum) - An in-depth look at the technology behind maglev trains and their potential for future transportation.

Online Resources

Khan Academy - Electromagnetism - A series of video lessons and interactive exercises on electromagnetism.

HyperPhysics - Magnetic Fields - An online resource with explanations and diagrams about magnetic fields and their interactions.

Journals and Periodicals

Journal of Applied Physics - Research articles on the latest developments in electromagnetism and its applications.

IEEE Transactions on Magnetics - Papers focused on the science and engineering of magnetic phenomena and devices.

Chapter 5

Alternating Current (AC) and Direct Current (DC)

5.1. Differences Between AC and DC

Understanding the differences between alternating current (AC) and direct current (DC) is crucial for anyone studying electrical engineering or working with electrical systems. As mentioned in Chapter 2, these two types of electrical current are fundamental to how electricity is generated, transmitted, and used in various applications. This section will explore the key differences between AC and DC, their respective advantages and disadvantages, and their common uses.

Nature of the Current Flow

- **Direct Current (DC)**: In a direct current (DC) system, the electric charge flows in a single, constant direction. The voltage in a DC circuit is steady and unchanging over time. This type of current is typically produced by sources such as batteries, solar cells, and DC generators. In DC circuits, electrons move from the negative terminal to the positive terminal of the power source.

- **Alternating Current (AC)**: In an alternating current (AC) system, the electric charge periodically reverses direction. This means that the voltage alternates between positive and negative values, creating a sinusoidal wave when graphed over time. AC is the standard form of electricity delivered to homes and businesses through the power grid. In AC circuits, electrons oscillate back and forth, rather than traveling in a single direction.

Voltage and Frequency

- **DC Voltage**: The voltage in a DC system remains constant, making it ideal for applications where a stable voltage is required. For example, electronic devices, such as laptops and smartphones, operate on DC power, which is why they use batteries or require AC adapters to convert AC to DC.

- **AC Voltage**: The voltage in an AC system varies over time, typically following a sinusoidal pattern. The frequency of this variation is measured in hertz (Hz), which indicates the number of cycles per second. In most countries, the standard frequency is either 50 Hz or 60 Hz, depending on the region. For instance, in the United States, the frequency is 60 Hz, while in Europe, it is 50 Hz.

Generation and Transmission

- **DC Generation**: DC power is generated by sources such as batteries, solar panels, and DC generators. It is suitable for short-distance transmission but is less efficient for long-distance power transmission due to higher energy losses over extended distances.

- **AC Generation**: AC power is generated by AC generators (alternators) and is the standard for power generation in most power plants. AC is more efficient for transmitting electricity over long distances because it can be easily transformed to higher or lower voltages using transformers, minimizing energy losses during transmission.

Conversion Between AC and DC

- **Rectification**: The process of converting AC to DC is called rectification. This is commonly done using devices known as rectifiers, which allow current to flow in only one direction. Rectifiers are used in power supplies for electronic devices, where AC from the wall outlet is converted to the DC needed by the device.

- **Inversion**: Converting DC to AC is known as inversion, performed by inverters. Inverters are used in applications such as uninterruptible power

Alternating Current (AC) and Direct Current (DC)

supplies (UPS) and solar power systems, where DC power from batteries or solar panels is converted to AC for use in standard electrical devices.

Applications

- **DC Applications**:

    o **Electronics**: DC is used to power electronic devices, including computers, smartphones, and other gadgets.

    o **Batteries**: All batteries provide DC power, making them essential for portable electronics and emergency power sources.

    o **Electric Vehicles**: DC is used in electric vehicle (EV) batteries, as well as in the motors of some EV designs.

- **AC Applications**:

    o **Power Distribution**: AC is used for distributing electricity from power plants to homes and businesses because it can be easily transformed to different voltage levels.

    o **Household Appliances**: Most household appliances, including refrigerators, washing machines, and air conditioners, are designed to run on AC power.

    o **Industrial Equipment**: AC is widely used in industrial machinery and equipment due to its efficiency in long-distance transmission and ease of voltage transformation.

Advantages and Disadvantages

- **Advantages of DC**:

    o Provides a constant voltage, which is necessary for electronic circuits.

    o Ideal for short-distance power transmission.

    o Simpler and safer to work with in low-voltage applications.

- **Disadvantages of DC**:
    - Less efficient for long-distance power transmission due to higher energy losses.
    - Requires more complex systems to convert from AC (which is typically supplied by the power grid).

- **Advantages of AC**:
    - More efficient for transmitting power over long distances due to the ability to use transformers.
    - Easier to generate and distribute on a large scale.
    - Compatible with most household and industrial devices.

- **Disadvantages of AC**:
    - Voltage varies, which can be less suitable for sensitive electronic devices without conversion to DC.
    - More complex to store in batteries, as they naturally produce DC.

5.2. Uses of AC and DC in Everyday Applications

Both alternating current (AC) and direct current (DC) are essential to the functioning of modern society, powering everything from household appliances to industrial machinery. Understanding where and how each type of current is used can provide valuable insights into the design and operation of electrical systems.

AC in Household Applications

- **Power Distribution**: AC is the standard form of electricity supplied to homes and businesses. The reason for this is its efficiency in transmission over long distances, as well as its ability to be easily transformed to different voltage levels using transformers. Power lines carry AC from power plants to substations, where the voltage is stepped down to a safer level before being delivered to homes.

- **Lighting**: Most household lighting systems, including incandescent bulbs, fluorescent lights, and LED lamps (when connected to standard fixtures), operate on AC power. The widespread availability of AC in homes makes it the most convenient option for lighting.

- **Appliances**: Major household appliances such as refrigerators, air conditioners, washing machines, and ovens are designed to operate on AC. These devices often require significant amounts of power, and AC's ability to be transformed to higher voltages makes it ideal for delivering the necessary energy efficiently.

- **HVAC Systems**: Heating, ventilation, and air conditioning (HVAC) systems rely on AC to power compressors, fans, and other components. The ability to efficiently transmit power over long distances and to easily control the voltage makes AC the preferred choice for these systems.

DC in Household Applications

- **Electronics**: Virtually all electronic devices, including smartphones, laptops, televisions, and gaming consoles, operate on DC power. These devices require stable and precise voltages, which DC provides. However, since household outlets supply AC, these devices use power adapters or chargers that contain rectifiers to convert AC to DC.

- **Batteries**: Batteries are a primary source of DC power in homes, providing portable energy for flashlights, remote controls, and other gadgets. Rechargeable batteries in laptops, smartphones, and other portable devices are charged using DC power, even though the charger typically plugs into an AC outlet.

- **Solar Panels**: Solar photovoltaic (PV) panels generate DC electricity when exposed to sunlight. This DC power can be used to charge batteries or, with the help of an inverter, converted to AC for use in the home or for feeding back into the electrical grid.

- **Low-Voltage Lighting**: Some home lighting systems, particularly those using LED strips or landscape lighting, operate on low-voltage DC. These systems are powered by a transformer or adapter that converts the AC from the outlet to the required DC voltage.

AC in Industrial Applications

- **Machinery**: Industrial machinery, such as conveyor belts, cranes, and manufacturing equipment, typically operates on AC power. The ability to deliver large amounts of power efficiently makes AC ideal for these heavy-duty applications.

- **Motors**: AC motors are widely used in industrial settings due to their robustness, efficiency, and ability to operate on the same AC power supply as the rest of the industrial infrastructure. These motors power everything from small tools to large industrial machines.

- **Power Transmission**: Large-scale power transmission networks rely on AC to transmit electricity over vast distances. High-voltage AC is used to minimize energy losses during transmission, and transformers step the voltage down to safer levels for distribution in factories and industrial plants.

- **Welding Equipment**: Many types of welding machines use AC power due to its availability and the ability to easily adjust the voltage and current for different welding tasks.

DC in Industrial Applications

- **Electroplating and Electrolysis**: DC power is used in electroplating processes to deposit a layer of metal onto a surface. It is also used in electrolysis, where electrical energy is used to drive a chemical reaction, such as the production of chlorine and sodium hydroxide.

- **Electric Vehicles**: DC power is used in the batteries and motors of electric vehicles (EVs). The batteries store energy as DC, and the motor controllers convert this energy into a form suitable for driving the vehicle. Although EVs are charged from AC outlets, the onboard charger converts this AC to DC to charge the batteries.

- **Backup Power Systems**: Many industrial facilities use DC in their uninterruptible power supplies (UPS) to provide emergency power during outages. These systems use batteries to store DC energy, which can be converted to AC by an inverter when needed.

- **Railway Systems**: Some railway systems, particularly those in urban environments, use DC power to drive trains. The DC power is often supplied by overhead lines or a third rail, providing consistent and reliable energy for traction motors.

AC vs. DC: Choosing the Right Current for the Job

The choice between AC and DC in any given application depends on several factors:

- **Distance**: AC is generally preferred for long-distance transmission due to its ability to be easily transformed to higher voltages, reducing energy loss. DC, however, is becoming more viable for long-distance transmission with the development of high-voltage direct current (HVDC) technology, which offers improved efficiency over very long distances.

- **Power Stability**: DC is favored in applications requiring stable and constant voltage, such as electronics and battery-operated devices. AC, with its fluctuating voltage, is more suitable for general power distribution and applications where voltage transformation is needed.

- **Efficiency**: In applications like electric vehicles and renewable energy systems, where energy efficiency is critical, DC is often used because it allows for more straightforward energy storage and conversion.

- **Safety and Simplicity**: For low-voltage applications, DC is often safer and easier to work with, particularly in portable and small-scale applications. AC is more complex to store and manage but excels in scenarios where variable voltage and long-distance transmission are necessary.

5.3. Introduction to Transformers

Transformers are a critical component in the transmission and distribution of electrical power, enabling the efficient transfer of electricity across long distances and the adjustment of voltage levels to meet different needs. Understanding how transformers work and their role in electrical systems is essential for anyone studying electrical engineering or working with electrical infrastructure.

What is a Transformer?

A transformer is an electrical device that transfers electrical energy between two or more circuits through electromagnetic induction. It works by converting electrical energy from one voltage level to another, either stepping up (increasing) or stepping down (decreasing) the voltage as needed. Transformers play a key role in ensuring that the electricity generated at power plants can be transmitted efficiently across the grid and then delivered at safe, usable voltage levels to homes, businesses, and industries.

Basic Principle of Operation

The operation of a transformer is based on the principles of electromagnetic induction, which were first discovered by Michael Faraday. A transformer typically consists of two coils of wire, known as the primary and secondary windings, which are wound around a common core made of ferromagnetic material, such as iron.

- **Primary Winding**: This coil is connected to the input power source, where an alternating current (AC) flows. As the AC flows through the primary winding, it creates a time-varying magnetic field around the coil.

- **Secondary Winding**: This coil is located near the primary winding and is influenced by the magnetic field generated by the primary winding. The changing magnetic field induces a voltage in the secondary winding, which is the output voltage of the transformer.

The voltage induced in the secondary winding is directly related to the number of turns of wire in each winding. This relationship is described by the transformer equation:

$$\frac{V_s}{V_p} = \frac{N_s}{N_p}$$

Where:

- V_s is the secondary voltage,
- V_p is the primary voltage,
- N_s is the number of turns in the secondary winding,
- N_p is the number of turns in the primary winding.

This equation shows that if the secondary winding has more turns than the primary winding, the transformer will step up the voltage. Conversely, if the secondary winding has fewer turns, the transformer will step down the voltage.

Types of Transformers

There are several types of transformers, each designed for specific applications:

- **Step-Up Transformers**:
  - **Purpose**: Increase the voltage from the primary to the secondary winding.
  - **Application**: Used in power generation stations to raise the voltage to very high levels for long-distance transmission. High voltage reduces energy losses over long distances.

- **Step-Down Transformers**:
  - **Purpose**: Decrease the voltage from the primary to the secondary winding.

- - Application: Used in substations and at the point of use to reduce the voltage to safe levels for homes, businesses, and industrial equipment.

- **Isolation Transformers**:
  - Purpose: Provide electrical isolation between the primary and secondary circuits without changing the voltage level.
  - Application: Used in sensitive electronic equipment to protect against electrical surges and interference.

- **Autotransformers**:
  - Purpose: Share part of the winding between the primary and secondary circuits, allowing for smaller, lighter transformers with adjustable voltage output.
  - Application: Used in applications requiring variable voltage control, such as in laboratory equipment and some industrial processes.

The Role of Transformers in Power Transmission

One of the most important uses of transformers is in the transmission and distribution of electrical power. In a typical power grid:

Power Generation: Electricity is generated at relatively low voltages (around 11kV to 33kV) in power plants. Step-up transformers increase this voltage to much higher levels (up to 765kV) for efficient long-distance transmission.

Transmission: The high-voltage electricity is transmitted over long distances through power lines. High voltage is crucial for reducing energy losses that occur due to the resistance of the transmission lines.

Substations: As electricity approaches populated areas, step-down transformers in substations reduce the voltage to intermediate levels (e.g., 33kV or 11kV) for distribution within the local network.

Distribution: Finally, at the point of use, such as in neighborhoods or industrial parks, another set of step-down transformers further reduces the voltage to the standard levels used by consumers (e.g., 120V/240V for homes in the U.S.).

Efficiency of Transformers

Transformers are highly efficient devices, with efficiency rates often exceeding 95%. This high efficiency is due to the low losses that occur primarily in the form of heat generated by the resistance of the windings and the hysteresis and eddy currents in the core material. Engineers continually work to improve transformer designs to minimize these losses and improve overall efficiency.

Applications of Transformers

Transformers are used in a wide range of applications beyond power transmission:

- **Electronics**: Transformers are found in power supplies for electronic devices, where they convert high-voltage AC from the outlet to the lower-voltage DC required by devices like laptops, smartphones, and televisions.

- **Industrial Machinery**: Large industrial machines often require specific voltage levels for operation. Transformers ensure that the appropriate voltage is delivered to these machines, regardless of the voltage provided by the grid.

- **Renewable Energy Systems**: In renewable energy systems, such as solar and wind power, transformers play a critical role in converting the variable voltage output of these systems to the standard voltage required for grid integration.

Alternating Current (AC) and Direct Current (DC)

5.4. Quiz for Review

Access Additional Quiz Material

Scan the QR code below to download a PDF file with additional quiz questions. This allows you to practice offline and further reinforce your understanding of the concepts covered in the book.

Alternating Current (AC) and Direct Current (DC)

5.5. Practical Exercises

Download Additional Exercises

Scan the QR code below to download a PDF file containing additional practical exercises. These exercises will help you apply the concepts learned in this book and further enhance your hands-on skills.

5.6. Curiosities and Further Insights

The concepts of alternating current (AC) and direct current (DC) are not only fundamental to electrical engineering but also play intriguing roles in various aspects of technology, history, and even the natural world. This section explores some of the more fascinating and lesser-known aspects of AC and DC, highlighting their impact and significance beyond the basics.

The War of the Currents: Edison vs. Tesla

One of the most famous historical battles in the field of electrical engineering is the "War of the Currents," which took place in the late 19th century. This was a conflict between two brilliant inventors, Thomas Edison and Nikola Tesla, over the preferred method of electrical power distribution.

- **Thomas Edison** advocated for direct current (DC) as the standard for power distribution. He believed that DC was safer and more reliable, and he had already established a number of DC power stations.

- **Nikola Tesla**, on the other hand, supported alternating current (AC), which he believed was more efficient for long-distance transmission. Tesla's AC system, backed by industrialist George Westinghouse, ultimately proved to be superior, especially after the successful demonstration of AC at the 1893 World's Columbian Exposition in Chicago.

The victory of AC led to its adoption as the standard for power generation and distribution worldwide, though DC continues to be used in many applications, particularly in electronics.

The Role of DC in Modern Electronics

While AC became the standard for power distribution, DC remains essential in the world of modern electronics. Almost all electronic devices, from smartphones to computers, operate on DC. This is because electronic circuits require a constant voltage to function properly, which DC provides.

- **Batteries**: All batteries supply DC power, making them crucial for portable electronic devices.

- **Power Adapters**: Devices that plug into an AC outlet often have power adapters (or "bricks") that convert AC to DC, providing the necessary voltage for the device.

Interesting Fact: The USB ports on your computer or charger provide 5V DC power, which is used to charge small devices like phones and tablets. The ubiquity of USB ports demonstrates the importance of DC in our daily lives.

AC in Nature: Lightning

One of the most powerful examples of alternating current in nature is lightning. During a thunderstorm, the movement of charged particles creates a massive electrical potential difference between clouds and the ground, leading to a sudden discharge in the form of lightning.

- **AC Characteristics**: The rapid oscillation of electric charge during a lightning strike results in a current that alternates, similar to man-made AC but on a much more intense and less controlled scale.

- **High Voltage**: Lightning can generate voltages of millions of volts and currents of tens of thousands of amperes, making it an extreme example of AC in nature.

Curiosity: The study of lightning and its effects has led to advancements in surge protection and grounding techniques, helping to protect electrical systems from the destructive power of these natural AC events.

The Development of HVDC Transmission

While AC is the standard for most power transmission, high-voltage direct current (HVDC) systems have been developed for specific long-distance and underwater transmission applications. HVDC offers certain advantages over AC in these scenarios, including lower energy losses and the ability to interconnect asynchronous grids.

- **Efficiency**: HVDC is more efficient for transmitting power over very long distances, especially when connecting power grids that operate on different frequencies or are located in different countries.

- **Application**: HVDC systems are used to transmit power between countries in Europe, across long distances in China, and through undersea cables in projects like the NorNed cable between Norway and the Netherlands.

Interesting Fact: The world's longest HVDC transmission line, the Belo Monte transmission line in Brazil, stretches over 2,500 kilometers and transmits 800 kV of power from the Belo Monte Dam to the southeastern region of the country.

The Future of DC in Renewable Energy

As renewable energy sources like solar and wind power become more prevalent, DC is gaining renewed importance. Solar panels, for example, generate DC power, which must be converted to AC for use in the grid. However, there is a growing interest in creating DC microgrids for more efficient energy use in certain applications.

- **DC Microgrids**: These are small-scale power grids that operate on DC and can be more efficient for integrating renewable energy sources, reducing the need for multiple conversions between AC and DC.

- **Energy Storage**: Battery storage systems, which are critical for managing the intermittent nature of renewable energy, naturally operate on DC, making them well-suited for integration into DC-based systems.

Curiosity: Some modern buildings, especially those focused on energy efficiency and sustainability, are experimenting with DC-based power systems to directly utilize the DC output from solar panels and batteries, minimizing energy losses associated with conversion.

The Versatility of AC Motors

AC motors are some of the most widely used machines in industry, found in everything from household appliances to large industrial machinery. Their ability to operate efficiently on the AC power grid and their relatively simple construction make them incredibly versatile.

- **Induction Motors**: One of the most common types of AC motors, induction motors, are used in everything from fans and refrigerators to conveyor belts and pumps. They operate by inducing a magnetic field in the rotor without requiring direct electrical connection to it.

Interesting Fact: Nikola Tesla invented the first practical AC induction motor in 1887, revolutionizing the use of electrical power and leading to the widespread adoption of AC in industry.

References and Further Reading for Chapter 5

Books

"AC/DC: The Savage Tale of the First Standards War" by Tom McNichol - A fascinating historical account of the battle between Edison and Tesla.

"Principles of Electric Circuits: Conventional Current Version" by Thomas L. Floyd - A comprehensive guide to understanding AC and DC circuits.

Articles and Publications

"The Evolution of HVDC Transmission Systems" (IEEE Spectrum) - An article discussing the development and advantages of HVDC technology.

"Lightning: Nature's High Voltage AC" (National Geographic) - An exploration of lightning as a natural phenomenon and its impact on electrical systems.

Online Resources

Khan Academy - Electricity and Magnetism - A series of educational videos and exercises on the basics of electricity, including AC and DC.

HyperPhysics - Alternating Current - An online resource with detailed explanations of AC concepts and applications.

Journals and Periodicals

Journal of Electrical Engineering and Technology - Research articles on the latest developments in AC and DC technologies.

IEEE Transactions on Power Delivery - Papers focused on the transmission and distribution of electrical power, including both AC and DC systems.

Chapter 6

Semiconductors and Electronic Components

Introduction to Semiconductors

Semiconductors are the foundation of modern electronics, playing a crucial role in the operation of virtually all electronic devices, from simple diodes to complex microprocessors. Understanding the basics of semiconductors is essential for anyone interested in electrical engineering or electronics, as these materials enable the control and manipulation of electrical signals in ways that have revolutionized technology.

What are Semiconductors?

Semiconductors are materials with electrical conductivity that falls between that of conductors (like copper) and insulators (like glass). This unique property makes semiconductors incredibly useful in electronic devices, as their conductivity can be easily manipulated by the introduction of impurities (a process known as doping) and by external factors such as temperature, light, and electric fields.

The most commonly used semiconductor material is silicon (Si silicium), due to its abundant availability and excellent electronic properties.

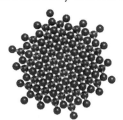 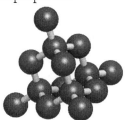

Other semiconductor materials include germanium and compounds like gallium arsenide (GaAs), which are used in specialized applications.

Properties of Semiconductors

Intrinsic Semiconductors:

Pure Form: An intrinsic semiconductor is a pure semiconductor material without any significant impurities. In this state, it has a relatively low conductivity because there are few charge carriers (electrons and holes) available to conduct electricity.

Behavior at Absolute Zero: At absolute zero temperature, an intrinsic semiconductor behaves almost like an insulator, as there are no free charge carriers. However, as the temperature increases, thermal energy causes electrons to jump from the valence band to the conduction band, creating electron-hole pairs that contribute to electrical conductivity.

Extrinsic Semiconductors:

Doping: The conductivity of a semiconductor can be dramatically increased by adding small amounts of impurities, a process known as doping. Doped semiconductors are classified into two types based on the type of impurity added:

- **N-type Semiconductors**: These are created by adding donor impurities (such as phosphorus) that provide extra electrons, increasing the number of negative charge carriers (electrons) in the material.
- **P-type Semiconductors**: These are created by adding acceptor impurities (such as boron) that create "holes" in the material, increasing the number of positive charge carriers (holes).
- **PN Junctions**: When N-type and P-type materials are brought together, they form a PN junction, which is the basis for many semiconductor devices such as diodes and transistors.

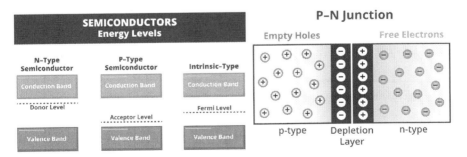

Why Are Semiconductors Important?

The unique ability of semiconductors to conduct electricity under certain conditions, while acting as insulators under others, makes them incredibly valuable for creating electronic components that can switch, amplify, and modulate electrical signals. This versatility is what allows semiconductors to be the building blocks of integrated circuits (ICs) and microchips, which are at the heart of all modern electronic devices.

- **Diodes**: A diode is a simple semiconductor device that allows current to flow in one direction only. It is made by joining N-type and P-type materials to form a PN junction. Diodes are used in rectification, signal demodulation, and many other applications.

- **Transistors**: Transistors are more complex devices that can act as switches or amplifiers. They are made by combining two PN junctions in either NPN or PNP configurations. Transistors are the fundamental building blocks of digital circuits, enabling everything from basic logic gates to complex microprocessors.

The Role of Semiconductors in Modern Electronics

Semiconductors have revolutionized the world of electronics by enabling the miniaturization and increased power of electronic devices. Before the development of semiconductor technology, electronic circuits relied on bulky vacuum tubes, which were less efficient, less reliable, and much larger.

With the advent of semiconductors, particularly silicon-based transistors, it became possible to create smaller, more efficient, and more reliable electronic devices. This led to the development of integrated circuits, where thousands or even millions of transistors are fabricated on a single silicon chip, allowing for the creation of powerful microprocessors and memory devices.

- **Computing**: Semiconductors are the foundation of all modern computing devices. Microprocessors, which are the brains of computers, are made up of billions of transistors on a single chip. The ability to pack more transistors into smaller spaces has driven the exponential growth in computing power over the past several decades, in line with Moore's Law.

- **Telecommunications**: Semiconductors are also critical in telecommunications, where they are used in everything from the chips in smartphones to the amplifiers and switches in communication networks.

- **Power Electronics**: In power electronics, semiconductors are used in devices such as rectifiers, inverters, and voltage regulators, which are essential for converting and controlling electrical energy in various applications, including renewable energy systems and electric vehicles.

6.1. Diodes, Transistors, and Their Uses

Semiconductors form the backbone of modern electronics, with diodes and transistors being two of the most essential components. These devices are crucial for various functions, such as switching, amplifying, and rectifying electrical signals. Understanding how diodes and transistors work, along with their applications, is key to mastering electronic circuits.

Diodes

What is a Diode?

A diode is a semiconductor device that allows current to flow in only one direction, acting as a one-way valve for electricity. Diodes are created by joining N-type and P-type semiconductor materials to form a PN junction. This junction permits current flow when the diode is forward biased (positive voltage applied to the P-type material) but blocks current when reverse biased (positive voltage applied to the N-type material).

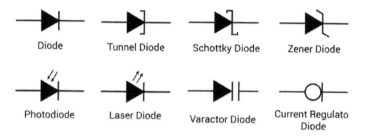

Types of Diodes:

- **Standard Diode**: Used primarily for rectification in power supplies, converting AC to DC. This process was discussed in the section on AC-DC conversion.

- **Zener Diode**: Conducts in reverse once a specific breakdown voltage is reached, making it ideal for voltage regulation.

- **Light-Emitting Diode (LED)**: Emits light when forward biased. LEDs are widely used in displays and lighting applications, which will be explored further in later chapters.

- **Schottky Diode**: Has a lower forward voltage drop, making it suitable for high-speed switching.

Applications of Diodes:

- **Rectification**: Essential in converting AC to DC, a fundamental process in power supplies.

- **Signal Demodulation**: Diodes are used to extract audio signals from modulated radio frequency signals.

- **Voltage Regulation**: Zener diodes are used in circuits to maintain a stable voltage.

- **Lighting**: LEDs have become the standard for energy-efficient lighting.

Transistors

What is a Transistor?

A transistor is a semiconductor device that can switch or amplify electrical signals. It consists of three layers of semiconductor material forming two PN junctions. The most common types are the bipolar junction transistor (BJT) and the field-effect transistor (FET).

- **Bipolar Junction Transistor (BJT)**: Has three terminals—emitter, base, and collector. An NPN transistor allows current to flow from the collector to the emitter when a small current is applied to the base. PNP transistors operate similarly but with reversed polarities.

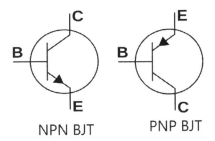

- **Field-Effect Transistor (FET)**: Has three terminals—source, gate, and drain. The voltage at the gate controls the current flowing from the source to the drain. FETs are widely used in digital circuits due to their efficiency and fast switching capabilities.

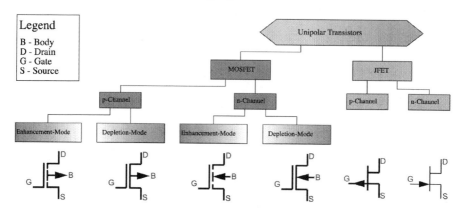

Types of Transistors:

- **NPN and PNP Transistors**: These are the two types of BJTs, each suited to different circuit designs.

- **Metal-Oxide-Semiconductor FET (MOSFET)**: Used in both analog and digital circuits, particularly in power electronics and microprocessors.

- **Darlington Transistor**: A configuration of two BJTs combined to provide high current gain.

Applications of Transistors:

- **Switching**: Transistors are the building blocks of digital logic circuits and are essential for binary logic operations in computers and other digital devices.

- **Amplification**: Transistors are used to amplify signals in a wide range of applications, from audio amplifiers in speakers to radio frequency amplifiers in communication devices.

- **Power Regulation**: In power supplies, transistors are used in voltage regulators and switching power supplies to control and stabilize the output voltage.

- **Oscillators**: Transistors are used in oscillator circuits to generate periodic signals, crucial for timing and frequency generation in electronic devices.

The Role of Diodes and Transistors in Integrated Circuits

Diodes and transistors are not only used as discrete components but are also integral parts of integrated circuits (ICs). In an IC, thousands or even millions of these components are fabricated on a single piece of semiconductor material, typically silicon, to perform complex functions.

- **Logic Gates**: The basic building blocks of digital circuits, constructed using transistors, perform simple logical operations like AND, OR, and NOT, forming the basis of all digital computing.

- **Microprocessors**: Modern microprocessors contain billions of transistors, allowing them to perform a wide range of tasks, from simple arithmetic to complex data processing.

- **Memory Devices**: Transistors are also used in memory chips, such as RAM and flash memory, where they store and retrieve data.

6.2. Advanced Overview of Essential Electronic Components

In previous chapters, we've introduced the basic electronic components such as resistors, capacitors, inductors, diodes, and transistors. This section will take a closer look at these components, exploring their advanced characteristics, specialized applications, and how they integrate into more complex electronic systems. This deeper understanding will prepare you for more sophisticated projects and circuit designs.

Advanced Resistor Applications

While resistors are fundamental components for limiting current and dividing voltage, their applications can extend far beyond simple circuits.

- **Precision Resistors**: Used in circuits where exact resistance values are crucial, such as in calibration equipment or high-precision measurement devices.

- **Temperature Coefficient**: Resistors are selected based on their temperature coefficient for environments where temperature changes can affect circuit performance. Thermistors, which were briefly mentioned earlier, are specifically designed to have a high temperature sensitivity, making them ideal for temperature sensing and compensation circuits.

- **Surface Mount Technology (SMT)**: In modern electronics, SMT resistors are used for their compact size and efficiency in automated circuit assembly, playing a crucial role in miniaturized devices like smartphones and laptops.

Capacitors in Advanced Circuitry

Capacitors are not only used for basic energy storage but are also critical in more complex applications.

- **High-Frequency Capacitors**: In RF (radio frequency) circuits, capacitors must handle high frequencies with minimal loss. Ceramic and mica capacitors are often used in these contexts for their stability and low loss at high frequencies.

- **Electrolytic Capacitors**: Commonly found in power supplies, these capacitors are crucial for smoothing voltage and reducing ripple in rectified DC circuits, especially in audio equipment and power amplifiers.

- **Supercapacitors**: With much higher capacitance than standard capacitors, supercapacitors are used in applications requiring rapid charge and discharge cycles, such as in regenerative braking systems in electric vehicles or as backup power in memory circuits.

Inductors in Advanced Applications

Inductors, which store energy in a magnetic field, are vital in various advanced electronic systems.

- **High-Frequency Inductors**: Used in RF circuits, these inductors must have low losses and maintain stability at high frequencies. Air-core inductors are preferred in these applications due to their lack of core losses.

- **Power Inductors**: In power supply circuits, inductors are used in DC-DC converters and switching regulators to smooth out voltage and filter noise. Their role is critical in maintaining power efficiency in devices ranging from computers to industrial machinery.

- **Transformers**: As a specialized form of inductors, transformers are crucial in AC power transmission and in converting voltages for different stages of electronic devices, as discussed in the section on AC and DC.

Diodes: Advanced Applications

Diodes, essential for rectification and circuit protection, also find applications in more sophisticated contexts.

- **Zener Diodes for Voltage Regulation**: These diodes are used to stabilize voltage in power supplies and sensitive circuits, keeping the output voltage constant despite variations in load or input voltage.

- **Schottky Diodes**: Known for their low forward voltage drop, these diodes are used in high-speed applications such as fast-switching rectifier circuits, improving energy efficiency in power converters.

- **LEDs and OLEDs**: Light-emitting diodes (LEDs) are widely used not only for lighting but also in flat-panel displays and mobile devices. Organic LEDs (OLEDs) represent an advanced evolution of LEDs, with applications in flexible displays and low-power screens.

Transistors in Complex Circuits

Transistors, vital for switching and amplification, are at the heart of both analog and digital electronics.

- **MOSFETs in Power Electronics**: Metal-Oxide-Semiconductor Field-Effect Transistors (MOSFETs) are widely used in power electronics for their efficiency in switching applications, particularly in DC-DC converters and motor control circuits.

- **BJTs in Analog Circuits**: Bipolar Junction Transistors (BJTs) are often used in analog circuits for amplifying weak signals in audio and radio frequency applications. Their ability to provide high gain makes them essential in designing amplifiers and oscillators.

- **Integrated Circuit Transistors**: In microprocessors, millions of transistors work together to perform complex calculations and process data. Understanding the role of individual transistors within an IC provides insight into the inner workings of modern computing devices.

Switches and Relays in Automation

Switches and relays are not just simple components for turning devices on or off; they are crucial in automation and control systems.

- **Solid-State Relays (SSRs)**: Unlike mechanical relays, SSRs have no moving parts and can switch currents much faster, making them ideal for industrial automation and precise control applications.

- **Microelectromechanical Systems (MEMS) Switches**: These tiny switches, used in advanced telecommunications and medical devices, offer high reliability and can operate at very high frequencies.

Specialized Components

Finally, a look at some specialized components that play key roles in advanced electronics:

- **Varistors**: Used for protecting circuits from voltage spikes, varistors are essential in surge protection devices, particularly in power strips and industrial equipment.

- **Quartz Crystals**: Used in oscillators to provide precise timing signals, quartz crystals are critical in communication devices, clocks, and microcontrollers.

- **Photodiodes and Phototransistors**: These components detect light and convert it into electrical signals, crucial in optical communication systems and light-sensitive devices.

6.3. Quiz for Review

Access Additional Quiz Material

Scan the QR code below to download a PDF file with additional quiz questions. This allows you to practice offline and further reinforce your understanding of the concepts covered in the book.

Semiconductors and Electronic Components

6.4. Practical Exercises

Download Additional Exercises

Scan the QR code below to download a PDF file containing additional practical exercises. These exercises will help you apply the concepts learned in this book and further enhance your hands-on skills.

6.5. Curiosities and Further Insights

Semiconductors and electronic components are not just the building blocks of modern technology; they also have fascinating histories, applications, and implications for the future. This section explores some of the more intriguing aspects of these components, highlighting their impact on the world and their potential to drive future innovations.

The Transistor's Impact on Modern Computing

The invention of the transistor in 1947 by John Bardeen, Walter Brattain, and William Shockley at Bell Labs revolutionized the world of electronics and computing. Before transistors, electronic devices relied on vacuum tubes, which were bulky, less reliable, and consumed a lot of power.

- **Miniaturization of Electronics**: Transistors allowed for the miniaturization of electronic devices, leading to the development of portable radios, hearing aids, and eventually, the modern computer.

- **Moore's Law**: Gordon Moore, co-founder of Intel, observed that the number of transistors on a microchip doubles approximately every two years, leading to exponential growth in computing power. This observation, known as Moore's Law, has driven the rapid advancement of technology for decades.

Interesting Fact: The first commercial computer, the UNIVAC I, used 5,000 vacuum tubes. In contrast, modern microprocessors contain billions of transistors, all on a chip no larger than a fingernail.

The Role of Semiconductors in Renewable Energy

Semiconductors play a crucial role in the development and implementation of renewable energy technologies. From solar panels to wind turbines, semiconductor components are at the heart of converting renewable sources into usable electricity.

- **Photovoltaic Cells**: Solar panels use semiconductor materials, typically silicon, to convert sunlight directly into electricity. When photons from sunlight strike the semiconductor material, they knock electrons loose, creating an electric current.

- **Power Converters**: Semiconductors are also used in power converters that manage the output from renewable energy sources, converting DC from solar panels into AC that can be used in homes or fed into the grid.

Interesting Fact: The largest solar power plant in the world, located in Bhadla, India, can generate over 2,245 megawatts of electricity, enough to power millions of homes. This incredible feat is made possible by the efficient conversion of sunlight into electricity using semiconductor technology.

The Evolution of LEDs: From Indicator Lights to Displays

Light-emitting diodes (LEDs) have come a long way from their initial use as simple indicator lights. Today, LEDs are used in everything from household lighting to massive digital displays and television screens.

- **Early LEDs**: The first LEDs, developed in the 1960s, emitted only red light and were used in low-power applications like indicator lights and seven-segment displays.

- **Modern Applications**: Advances in semiconductor materials have led to LEDs that can emit light in a wide range of colors, including white. These LEDs are now used in energy-efficient lighting, HD displays, and even automotive headlights.

Interesting Fact: The development of blue LEDs in the 1990s, for which the inventors won the Nobel Prize in Physics in 2014, made it possible to create white LEDs by combining red, green, and blue light. This innovation has had a significant impact on reducing global energy consumption.

Semiconductor Fabrication: A Modern Marvel

The process of fabricating semiconductor devices, such as microchips, is one of the most complex and precise manufacturing processes in the world.

- **Cleanrooms**: Semiconductor fabrication occurs in cleanrooms that are 100 times cleaner than a hospital operating room. Even a single speck of dust can destroy a microchip during production.

- **Photolithography**: One of the key processes in chip fabrication is photolithography, where a pattern is etched onto the semiconductor material using light. This process is repeated multiple times to create the intricate circuits that make up a microchip.

Interesting Fact: The smallest features on modern microchips are now measured in nanometers (billionths of a meter). The latest technology nodes, such as 5nm and 3nm, allow billions of transistors to be packed onto a single chip.

The Future of Semiconductors: Beyond Silicon

While silicon has been the dominant material in semiconductor devices for decades, researchers are exploring new materials that could lead to even more powerful and efficient electronics.

- **Graphene**: A single layer of carbon atoms arranged in a hexagonal lattice, graphene has remarkable electrical properties, including high conductivity and flexibility. It is being researched for use in transistors, sensors, and even flexible displays.

- **Gallium Nitride (GaN)**: GaN is a wide-bandgap semiconductor material that can handle higher voltages and temperatures than silicon, making it ideal for power electronics and RF applications. It is already being used in high-efficiency power converters and 5G technology.

Interesting Fact: Quantum computing, which promises to revolutionize computing by performing complex calculations much faster than classical computers, relies on qubits made from superconducting materials and other advanced semiconductors.

References and Further Reading for Chapter 6

Books

"The Chip: How Two Americans Invented the Microchip and Launched a Revolution" by T.R. Reid - *A detailed history of the invention of the microchip and its impact on the world.*

"Semiconductor Fundamentals" by Robert F. Pierret - *A comprehensive textbook covering the fundamental principles of semiconductor devices.*

Articles and Publications

"Moore's Law and the Future of Electronics" (IEEE Spectrum) - *An article discussing the continuing relevance of Moore's Law and its implications for the future.*

Online Resources

Khan Academy - Semiconductor Physics - *An educational resource offering video lessons on the basics of semiconductor physics.*

HowStuffWorks - How Transistors Work - *An accessible guide to understanding the operation and applications of transistors.*

Journals and Periodicals

Journal of Applied Physics - *Research articles on the latest advancements in semiconductor technology and applications.*

IEEE Transactions on Electron Devices - *Papers focused on the development and application of electronic devices, including semiconductors.*

Chapter 7

Practical Applications and Projects in Electrical Engineering

Introduction to Practical Applications

Electrical engineering is a vast and dynamic field that impacts nearly every aspect of modern life. From the devices we use daily to the infrastructure that powers our homes and cities, electrical engineering plays a crucial role in making our world more efficient, connected, and sustainable. Understanding the practical applications of electrical engineering not only enhances theoretical knowledge but also provides the hands-on experience necessary to innovate and solve real-world problems.

The Importance of Practical Applications

Practical applications in electrical engineering are where theory meets reality. They bridge the gap between conceptual understanding and functional design, allowing engineers to apply principles learned in the classroom to create tangible solutions. Whether you are developing a new consumer electronic device, designing a renewable energy system, or creating an intelligent home automation network, practical applications are key to advancing technology and improving quality of life.

- **Real-World Problem Solving**: Practical applications allow engineers to address specific challenges, such as increasing energy efficiency, enhancing communication systems, or improving safety. These applications require a deep understanding of both the underlying principles and the practical constraints of engineering projects.

- **Innovation and Creativity**: Working on practical projects fosters innovation by encouraging engineers to think creatively and explore new ways to apply existing technologies. This creative problem-solving approach often leads to breakthroughs that can have a significant impact on industries and society.

- **Adapting to Emerging Technologies**: In a rapidly evolving technological environment, electrical engineering increasingly integrates with emerging technologies like the Internet of Things (IoT), artificial intelligence, and robotics. Practical projects enable engineers to experiment with these new technologies, helping them understand how they can be integrated into existing systems to create advanced solutions. For example, IoT-based home automation systems demonstrate how traditional electrical design can evolve to embrace modern connectivity.

The Role of Practical Applications in Developing Future Technologies

Beyond solving current problems, practical applications allow engineers to explore technologies that could become fundamental in the near future. Advanced systems such as smart grids, renewable energy technologies, and electric vehicles represent areas where electrical engineering will play an increasingly crucial role. Electrical engineers must be prepared to innovate through practical applications that equip them for these challenges. This chapter will provide examples of how these developments can be applied in reality, paving the way for new growth opportunities.

The Role of Hands-on Projects in Learning and Innovation

Hands-on projects are integral to the learning process in electrical engineering. They provide an opportunity to experiment, make mistakes, and learn from those mistakes in a controlled environment. By engaging in practical projects, you can deepen your understanding of complex concepts, develop critical thinking skills, and gain the confidence needed to tackle more advanced challenges.

- **Learning by Doing**: There is no substitute for hands-on experience. Building circuits, testing components, and troubleshooting issues are essential activities that reinforce theoretical knowledge. These projects give you a chance to apply what you've learned in previous chapters, such as the principles of semiconductors, circuit design, and power management.

- **Connecting Theory with Practice**: Practical projects help you see how abstract concepts are applied in real-world scenarios. For example, understanding the role of transistors in amplifiers (as discussed in Chapter 6) becomes much clearer when you actually build and test a working amplifier circuit.

Overview of This Chapter

In this chapter, you will explore various practical applications of electrical engineering, ranging from consumer electronics to renewable energy systems. Each section will introduce a specific area of application, followed by a detailed explanation of how certain practical examples are realized in the field and how these align with electrical engineering concepts. These explanations are designed to provide valuable insights into the engineering principles behind each application, deepening your theoretical understanding and preparing you to apply similar approaches in your own designs.

- **Consumer Electronics**: Explore how electrical engineering powers devices like TVs, computers, and smartphones. We will explain how a simple audio amplifier for a smartphone is designed and the engineering concepts involved in its operation.

- **Home Automation and IoT**: Learn how electrical engineering integrates with smart home technologies. You will gain an understanding of how basic home automation systems are created using IoT devices, focusing on the interaction of sensors and controllers.

- **Renewable Energy Systems**: Delve into the world of solar panels and wind turbines. We will explain the construction of a small-scale solar power system and highlight the key components and design principles necessary for sustainable energy production.

- **Security and Monitoring Systems**: Discover the role of electrical engineering in home and industrial security. This section will cover how functional home alarm systems are engineered and the sensors involved.

- **Lighting Systems**: Study the evolution of lighting from incandescent bulbs to LEDs. We will explore how custom LED lighting systems are designed to maximize efficiency and light quality.

- **Robotics and Control Systems**: See how electrical engineering intersects with robotics. This section will cover how a simple remote-controlled robot is constructed, including motor control and sensor integration.

- **Mobile and Portable Power Solutions**: Learn about battery and solar-powered devices. We will explain the design behind a solar charger for mobile devices, focusing on power management and energy conversion techniques.

By the end of this chapter, you will have gained a deeper understanding of how electrical engineering is applied in real-world scenarios, with detailed examples that illustrate the connection between theory and practice. These explanations will strengthen your ability to design, troubleshoot, and innovate in various areas of electrical engineering.

Connecting with Previous Chapters

Throughout this chapter, you'll see references to concepts covered in earlier chapters. For instance, when working on the solar power project, you might revisit the discussion on semiconductors and photovoltaic cells in Chapter 6. Similarly, the design of the home automation system will draw upon your understanding of basic circuits and component functions, as discussed in Chapters 2 and 3. These connections will help reinforce your learning and ensure that you have a comprehensive understanding of the material.

Transition to Next Section

Now that you have an overview of the practical applications of electrical engineering and their relevance, it's time to explore the first area: consumer electronics. In the next section, you will discover how electrical engineering powers everyday devices, and we will explain the principles behind designing an audio amplifier for a smartphone, highlighting its integration with key concepts from previous chapters.

7.1. Consumer Electronics

Consumer electronics are an integral part of modern life, from the smartphones in our pockets to the televisions in our living rooms. Electrical engineering is at the core of these devices, enabling the functionality, reliability, and efficiency that we often take for granted. This section will explore how the principles of electrical engineering are applied in consumer electronics, focusing on key components and technologies that power these everyday devices.

The Role of Electrical Engineering in Consumer Electronics

Electrical engineering is essential in the design, development, and manufacturing of consumer electronics. Engineers in this field address various challenges, including power management, signal processing, communication systems, and user interfaces. Understanding how these components work together allows engineers to create devices that are not only functional but also user-friendly and energy-efficient.

- **Power Management**: Managing power consumption is critical in portable devices like smartphones and laptops. Engineers design circuits that maximize battery life while ensuring the device operates efficiently under various conditions.

- **Signal Processing**: Signal processing plays a crucial role in devices like televisions and audio equipment. Engineers develop algorithms and circuits to convert, compress, and enhance signals, ensuring high-quality audio and video output.

- **Communication Systems**: Consumer electronics often require communication with other devices or networks. Engineers design circuits and protocols that enable wireless communication, including Wi-Fi, Bluetooth, and cellular networks.

Key Components in Consumer Electronics

Consumer electronics rely on a variety of electronic components, many of which have been covered in earlier chapters. This section focuses on how these components are specifically used in consumer devices:

- **Microprocessors and Microcontrollers**: These are the brains of most modern electronics. From the central processing unit (CPU) in a computer to the microcontroller in a microwave oven, these components execute the software that drives the device and ensures functionality.

- **Display Technologies**: LCD, OLED, and LED displays are common in consumer electronics. Engineers design circuits to control these displays, ensuring clear, vibrant visuals. The development of LED technology has significantly improved the energy efficiency of displays.

- **Batteries and Power Supplies**: Batteries are essential for portable electronics, while power supplies convert AC to DC to power stationary devices. Engineers design efficient charging circuits and ensure safe operation, especially in devices like laptops and smartphones.

7.2. Home Automation and IoT (Internet of Things)

The integration of electrical engineering with modern technology has given rise to the Internet of Things (IoT), a network of interconnected devices that communicate with each other and the cloud to automate and optimize various aspects of daily life. Home automation, powered by IoT, exemplifies how electrical engineering enhances convenience, security, and energy efficiency in the home. This section explores the basics of home automation, the role of IoT, and how these technologies work together to create smart homes.

The Evolution of Home Automation

Home automation has progressed from basic, standalone devices like remote-controlled lights to highly sophisticated systems that manage entire homes. IoT technology has revolutionized the field, allowing devices to connect to the internet, communicate with each other, and provide a seamless, integrated experience for homeowners.

- **Early Home Automation**: Early systems were limited to simple tasks, such as turning lights on and off or adjusting thermostats. These systems lacked the ability to interact with other devices, functioning primarily as isolated systems.

- **Modern IoT-Enabled Homes**: Today's smart homes are equipped with IoT devices that can be controlled remotely via smartphones or voice assistants like Alexa and Google Home. These devices monitor and control lighting, heating, security systems, and appliances, creating a fully integrated and automated living environment.

Key Components of a Smart Home System

A smart home system consists of several interconnected devices, each performing a specific function. Here are some of the core components typically found in these systems:

- **Smart Hubs and Controllers**: The central hub or controller acts as the brain of the smart home system. It manages communication between devices and allows the user to control the system via an app or voice commands. Popular smart hubs include Amazon Echo, Google Nest Hub, and Apple HomePod.

- **Smart Lighting**: IoT-enabled lights can be remotely controlled, scheduled to turn on or off at specific times, or adjusted based on ambient light levels. Smart bulbs and switches are commonly used for this purpose.

- **Smart Thermostats**: Devices like the Nest Thermostat learn your schedule and preferences, adjusting the temperature automatically to optimize comfort and energy efficiency.

- **Smart Security Systems**: These systems include cameras, door locks, motion sensors, and alarms that can be monitored and controlled remotely. They provide real-time alerts and can integrate with other devices for enhanced security.

- **Smart Appliances**: From refrigerators that monitor food inventory to washing machines that can be started remotely, smart appliances have become standard features in modern homes, adding convenience and efficiency to everyday tasks.

How a Basic Smart Home System Works

The smart hub functions as the central control point for all connected devices, communicating over the home's Wi-Fi network. Through an app, users can set automation rules and schedules, allowing devices to operate independently based on preferences, such as turning on lights at sunset or adjusting the thermostat when leaving the house. Voice commands also enable seamless interaction with the system.

Troubleshooting Common Issues

- **Connectivity Problems**: Devices must be connected to the same Wi-Fi network, and the signal strength must be sufficient for communication. Rebooting the router or smart hub can sometimes resolve issues.

- **Device Pairing Issues**: Devices may occasionally fail to pair with the smart hub. Resetting the devices and ensuring they are in pairing mode, along with using the correct app, can resolve this.

- **Automation Failures**: Double-checking the automation rules and ensuring that devices are properly connected and responsive is essential for the smooth operation of a smart home system.

Applications of Smart Home Automation

The basic framework of a smart home system can be expanded to include a wide range of devices and systems. In real-world applications, smart home automation can:

- **Enhance Energy Efficiency**: By automating lighting, heating, and cooling, smart homes can reduce energy consumption and lower utility bills.

- **Improve Security**: Smart security systems provide real-time monitoring and alerts, offering improved protection for homes and families.

- **Increase Convenience**: Remote control and automation of household devices simplify daily routines, providing a more comfortable and streamlined living experience.

7.3. Renewable Energy Systems

Renewable energy is at the forefront of modern engineering, addressing the global need for sustainable and environmentally friendly power sources. Electrical engineering plays a pivotal role in capturing renewable energy from sources such as the sun and wind, converting it into electricity, and integrating it into the power grid. This section explores the fundamental principles of renewable energy systems, focusing on the critical components and technologies that make solar and wind power systems function efficiently.

The Role of Electrical Engineering in Renewable Energy

Electrical engineering is essential in the design, optimization, and deployment of renewable energy systems. Engineers in this field create and refine systems that convert natural energy sources into usable electrical power, while ensuring efficiency, safety, and reliability.

- **Solar Power**: Solar panels, or photovoltaic (PV) cells, convert sunlight directly into electricity. Engineers design circuits and systems to manage the flow of electricity from these panels, optimizing energy output and ensuring stability in variable conditions.

- **Wind Power**: Wind turbines harness kinetic energy from the wind and convert it into electrical energy. Electrical engineers develop the control systems that regulate turbine output and safely integrate it with the electrical grid.

- **Energy Storage**: Since renewable energy sources like solar and wind are intermittent, energy storage systems (such as batteries) are critical for providing a continuous power supply. Engineers work on efficient storage solutions and design systems that regulate energy flow between generation, storage, and consumption.

Key Components of Renewable Energy Systems

Renewable energy systems rely on several key components that work together to generate, convert, and store electricity:

- **Photovoltaic (PV) Cells**: These are the fundamental units in solar panels that convert sunlight into electricity. PV cells are made of semiconductor materials like silicon and generate DC electricity when exposed to sunlight.

- **Inverters**: Inverters convert the DC electricity generated by solar panels into AC electricity, which can be used to power household appliances or fed into the power grid.

- **Charge Controllers**: Charge controllers regulate the flow of electricity from solar panels to batteries, preventing overcharging and ensuring the battery operates efficiently over time.

- **Batteries**: Batteries store electricity generated by renewable sources for later use, especially when sunlight or wind is not available.

- **Wind Turbines**: Wind turbines capture kinetic energy from the wind and convert it into mechanical energy, which is then converted into electrical energy by a generator.

How Renewable Energy Systems Work

In a solar energy system, sunlight hits the PV cells, generating a DC voltage. This electricity is sent to a charge controller, which regulates the current before sending it to a battery for storage. If the system is connected to an inverter, the DC electricity can be converted to AC for household use or distribution to the grid. Wind energy systems follow a similar process, where wind drives the turbine to produce electricity, which is then regulated, stored, or used.

Common Issues and Solutions in Renewable Energy Systems

- **Low Power Output**: Positioning is key. Solar panels need to be placed where they receive maximum sunlight, while wind turbines must be installed in locations with consistent wind.

- **Battery Not Charging**: Ensure the charge controller is functioning properly, and check that the battery is in good condition and wired correctly.

- **Inconsistent Power Supply**: This can be due to fluctuating weather conditions or insufficient storage capacity. Expanding the system with more solar panels, wind turbines, or larger battery banks may help.

Real-World Applications of Renewable Energy Systems

Understanding how small-scale systems function is crucial for expanding their use in real-world applications. These systems are scalable and adaptable to various energy needs:

- **Residential Solar Power**: Homeowners can install solar panels to generate electricity, reduce their reliance on the grid, and lower energy bills. Larger systems with more PV cells and higher-capacity batteries can power entire homes.

- **Off-Grid Systems**: In remote areas, where access to the power grid is limited or unavailable, renewable energy systems provide a vital source of electricity. Solar and wind systems can power essential equipment, such as communication devices and refrigeration.

- **Sustainable Development**: Renewable energy systems are key to reducing carbon emissions and combating climate change. Large-scale solar farms and wind farms are increasingly being used to power entire communities and cities.

7.4. Security and Monitoring Systems

Security and monitoring systems are essential in both residential and industrial environments, offering protection by deterring intrusions, monitoring activity, and alerting users to potential threats. Electrical engineering plays a fundamental role in designing and implementing these systems, which range from basic door alarms to advanced surveillance networks. In this section, we'll explore the core components of security systems, the technology that supports them, and the underlying principles of their operation.

The Role of Electrical Engineering in Security Systems

Security systems combine sensors, cameras, alarms, and control panels to create a robust protection network. Electrical engineers develop these systems to be responsive, reliable, and user-friendly, ensuring they work effectively under various conditions.

- **Sensors**: Sensors detect environmental changes, such as motion, temperature, or the opening of doors and windows. Common sensor types include infrared (IR) sensors, magnetic reed switches, and glass break detectors.

- **Cameras**: Surveillance cameras monitor and record activities in key areas. Modern security systems often use IP cameras that connect to a network, enabling remote viewing and storage of footage.

- **Alarms**: When a sensor is triggered, the system activates an alarm. Alarms can be audible, visual (flashing lights), or silent (e.g., sending an alert to a smartphone or security personnel).

- **Control Panels**: The control panel is the system's "brain," receiving signals from sensors and cameras and determining the appropriate response. Most control panels are networked for remote access and monitoring.

Key Components of Security and Monitoring Systems

A typical security system is made up of multiple elements working together to detect and respond to threats:

- **Motion Sensors**: These sensors detect movement in a designated area. They are commonly used both indoors and outdoors to detect unauthorized entry.

- **Magnetic Reed Switches**: Installed on doors and windows, these switches detect when the entry point is opened or closed. They are integral to perimeter security.

- **Surveillance Cameras**: Cameras provide real-time monitoring and can record activity. They may operate continuously or be triggered by motion sensors.

- **Alarm Sirens**: Sirens are activated when the system detects an intrusion, producing loud sounds to scare off intruders and alert neighbors or security personnel.

- **Control Panels and Keypads**: These allow users to arm or disarm the system, monitor sensor status, and respond to alerts. Many systems now integrate with smartphones for remote access and control.

How Security Systems Work

Security systems rely on a network of interconnected components. When a sensor detects an abnormality (like a door opening or motion in a restricted area), it sends a signal to the control panel. The control panel interprets this signal and activates the appropriate response, such as sounding an alarm or sending a notification. Engineers design these systems to be reliable and responsive, ensuring quick action when needed.

Common Troubleshooting Issues

- **Alarm Not Sounding**: This may result from a misaligned reed switch or weak sensor signal. Ensure proper installation and connections between components.

- **False Alarms**: Common causes include faulty sensors or loose wiring. Regular maintenance and system checks can prevent false alarms.

- **Weak Alarm Sound**: If the alarm volume is low, the battery may need replacement, or resistors in the circuit may need adjustment.

Real-World Applications of Security Systems

Understanding the basic principles of security systems allows for insights into more advanced and integrated solutions used in both residential and commercial settings:

- **Residential Security**: Homeowners use systems with motion sensors, cameras, and alarms to secure entry points and monitor outdoor areas.

- **Commercial and Industrial Security**: In larger settings, security systems include advanced access control, high-resolution cameras, and specialized sensors to protect assets and personnel.

- **IoT Integration**: Modern security systems increasingly incorporate Internet of Things (IoT) devices, allowing remote monitoring and control through smartphone apps or other smart home platforms.

7.5. Lighting Systems

Lighting is one of the most fundamental applications of electrical engineering, with a rich history that spans from the invention of the incandescent light bulb to the development of energy-efficient LED technology. Today's lighting systems not only provide illumination but also enhance energy efficiency, sustainability, and aesthetics. This section explores the evolution of lighting technologies and the engineering principles that underlie modern LED systems.

The Evolution of Lighting Systems

The journey of lighting technology is a testament to the significant advances in electrical engineering, with each stage contributing to greater efficiency, longevity, and versatility.

- **Incandescent Bulbs**: Invented by Thomas Edison in 1879, incandescent bulbs produce light by heating a tungsten filament until it glows. As discussed in **Chapter 1**, Edison's contribution to electrical engineering marked a turning point in the development of practical applications for electricity. However, incandescent bulbs are inefficient, converting only about 10% of the energy into visible light, with the remainder lost as heat.

- **Fluorescent Lighting**: Fluorescent lamps, introduced in the mid-20th century, excite mercury vapor to produce ultraviolet light, which in turn causes a phosphor coating inside the bulb to glow. These lamps are more energy-efficient than incandescent bulbs, but their reliance on mercury poses environmental challenges.

- **Compact Fluorescent Lamps (CFLs)**: A smaller and more efficient version of fluorescent lamps, CFLs became popular as energy-efficient alternatives to incandescent bulbs. However, like their larger counterparts, CFLs still contain mercury, making disposal a concern.

- **Light Emitting Diodes (LEDs)**: The most significant advancement in lighting technology, LEDs generate light through electroluminescence, where a semiconductor material emits light when an electric current passes through it. LEDs are highly efficient, long-lasting, and versatile, making them the preferred choice for modern lighting solutions.

Key Components of LED Lighting Systems

LED lighting systems have become dominant due to their efficiency, longevity, and flexibility. Understanding the core components of an LED system allows engineers to optimize lighting for different applications.

- **LEDs**: The core components, made from semiconductor materials, convert a significant portion of electrical energy into light, with minimal heat production. This makes them far more efficient than older lighting technologies.

- **Drivers**: LED drivers regulate the voltage and current supplied to the LEDs, ensuring safe and efficient operation. Drivers convert incoming AC power to the DC power required by LEDs.

- **Heat Sinks**: Although LEDs generate less heat than traditional incandescent bulbs, they still produce some heat. Heat sinks help dissipate this heat away from the LEDs to prevent overheating and ensure longevity.

- **Diffusers and Lenses**: These components help distribute light evenly and reduce glare. Diffusers soften the light, while lenses can focus or spread the light as needed for various applications.

How It Works: The Engineering Behind LED Lighting

Understanding how LED lighting systems operate helps grasp the broader engineering principles involved:

- **Power Conversion**: LED drivers convert AC to DC and regulate the power provided to the LEDs, ensuring the correct voltage and current. This conversion is crucial to LED efficiency and longevity.

- **Heat Management**: Managing heat is vital in LED systems. The heat sink absorbs and dissipates the heat generated by the LEDs. Without proper heat management, LEDs can overheat, which reduces their lifespan.

- **Light Control**: Diffusers and lenses are used to control the distribution of light. Diffusers spread light more evenly, creating a soft, pleasant glow, while lenses can focus the light for more directed illumination. Engineers must carefully choose these components depending on the application—whether it's for home lighting or industrial use.

Practical Learning: Insights from LED Lighting Design

Rather than focusing on physically constructing a system, readers should understand the thought process behind designing an efficient LED lighting system:

- **Design Considerations**: Engineers must consider factors such as the desired brightness, light distribution, and energy consumption when designing LED systems. This includes selecting the appropriate LEDs, drivers, and heat sinks for the specific application.

- **Energy Efficiency**: LED systems are highly efficient compared to older technologies, largely due to their ability to convert most electrical energy into light. Understanding this efficiency helps engineers design systems that meet modern energy standards.

- **Sustainability**: The longevity of LEDs, combined with their lower energy consumption, makes them a sustainable option for reducing carbon footprints. This knowledge is critical for engineers working in industries focused on green technologies and sustainability.

Applications in Real-World Lighting Systems

The engineering principles behind LED lighting translate directly into real-world applications, including:

- **Residential Lighting**: LEDs are frequently used in homes for efficient, long-lasting lighting solutions, reducing household energy consumption. Engineers designing residential lighting systems must balance efficiency with aesthetics, choosing appropriate diffusers and heat sinks for compact, stylish fixtures.

- **Commercial Lighting**: In offices and retail environments, LED lighting is preferred for its low maintenance and energy savings. Engineers focus on maximizing brightness while minimizing energy costs, often incorporating smart controls to optimize usage.

- **Automotive Lighting**: LEDs are now standard in automotive headlights, taillights, and interior lighting. Their durability and brightness make them ideal for automotive use, where reliability and visibility are critical. Engineers in this field design systems that can withstand the harsh conditions of automotive environments, including temperature fluctuations and vibrations.

7.6. Robotics and Control Systems

Robotics represents one of the most exciting and rapidly advancing fields in electrical engineering. It combines principles from electronics, mechanics, and computer science to create machines that can perform tasks autonomously or under human control. Control systems are integral to robotics, allowing precise management of movement, decision-making, and interaction with the environment. In this section, we will explore the basics of robotics and control systems and focus on how electrical engineering integrates into these systems.

The Intersection of Electrical Engineering and Robotics

Electrical engineering plays a central role in robotics, providing the necessary power, control, and sensing capabilities that allow robots to function. Robots can range from simple mechanical devices to highly complex systems capable of sophisticated tasks. Understanding the core components and how they interact is key to the design of functional robotic systems.

- **Power Systems**: Robots require power to operate their motors, sensors, and control systems. Electrical engineers design power systems that provide the necessary energy, whether sourced from batteries, solar panels, or other energy systems.

- **Control Systems**: Control systems manage the operation of a robot, including its movements, responses to sensors, and execution of programmed tasks. These systems can range from simple manual controls to complex algorithms that allow for autonomous behavior, such as in self-driving cars or robotic arms in industrial applications.

- **Sensors and Actuators**: Sensors allow robots to perceive their environment, while actuators convert electrical signals into mechanical movement. Together, they enable robots to interact with their surroundings. For example, ultrasonic sensors help robots detect obstacles, while actuators control the movement of robotic arms.

Key Components of a Robot

A basic robot consists of several key components that work together to perform tasks:

- **Microcontroller**: Often referred to as the brain of the robot, the microcontroller processes inputs from sensors, makes decisions, and controls actuators. Common microcontrollers used in robotics include platforms like Arduino and Raspberry Pi.

- **Motors and Servos**: Motors provide the movement needed for wheels, arms, and other parts of the robot. Servos are specialized motors that allow for precise control of angular position, which is essential in robotic arms and steering mechanisms.

- **Sensors**: Sensors collect data about the robot's environment. For example, ultrasonic sensors detect distance, light sensors measure ambient light, and gyroscopes track orientation. The information from these sensors is used to make decisions about the robot's actions.

- **Chassis**: The chassis is the physical frame that holds all the robot's components together. It is typically designed to withstand the demands of the robot's environment and application.

- **Communication Modules**: In systems that require remote control, communication modules such as RF (radio frequency) transmitters and receivers allow for wireless communication between the operator and the robot.

Understanding the Design of a Simple Robot

To understand how robots are designed, it is helpful to break down the design process into its components and functions, focusing on the theoretical aspects:

- **Chassis and Structural Design**: The design begins with selecting a chassis that suits the application. For example, a wheeled chassis might be ideal for a robot navigating flat surfaces, while a tracked chassis may be more appropriate for rough terrain.

- **Motor Control**: Motors are chosen based on the type of movement required. Engineers calculate the power needed to drive the motors, taking into account factors such as weight, friction, and the desired speed of the robot. The motor driver circuit then controls the motors, adjusting voltage and current as needed.

- **Programming the Microcontroller**: The microcontroller is programmed to manage the robot's movement and reactions to sensor data. For instance, it can be programmed to stop the robot when an obstacle is detected, or to follow a pre-defined path. This process involves writing code that defines how the robot will interpret sensor data and translate it into movement or other actions.

How Control Systems Work in Robotics

Control systems are vital to robotics as they regulate the movement and actions of the robot based on sensor inputs and programmed algorithms. Here's how the main control processes work:

- **Feedback Control**: Feedback control systems continuously monitor a robot's environment through sensors and adjust the robot's actions accordingly. For example, a temperature sensor in a robotic arm might trigger cooling mechanisms if the arm begins to overheat during operation.

- **Open-Loop vs. Closed-Loop Systems**: Open-loop control systems execute a set of instructions without adjusting to environmental feedback, such as a simple conveyor belt that runs at a constant speed. Closed-loop systems, on the other hand, modify their actions based on sensor feedback, allowing for more precise control, as in the case of a self-balancing robot.

Applications in Real-World Robotics

The engineering principles that underpin robotics are applied in various real-world contexts:

- **Autonomous Robots**: Autonomous robots, such as those used in warehouses for logistics, rely on advanced sensors, control systems, and algorithms to navigate and perform tasks without human intervention.

- **Industrial Automation**: Robots in manufacturing environments perform repetitive tasks like assembly, welding, and material handling. These robots need precise control and robust power systems to ensure reliability and safety in their operations.

- **Educational Robots**: Many educational robotics kits are designed to teach students the basics of robotics, including how to program microcontrollers and control motors. These kits provide an accessible way to explore the core concepts of robotics and control systems.

Learning Outcomes from Robotic Systems Design

Studying the design and operation of robotic systems provides valuable insights into several areas of electrical engineering, including:

- **Power Management**: Understanding how to efficiently power robotic systems is crucial, particularly when dealing with mobile robots that rely on batteries. Engineers must consider energy efficiency and longevity in their designs.

- **Control Theory**: Mastering control theory allows engineers to create more precise and reliable systems. Control systems are used not only in robotics but also in other fields such as aerospace, automotive engineering, and even consumer electronics.

- **Sensor Integration**: Proper integration of sensors is essential for a robot's ability to interact with its environment. Engineers must select the right sensors for the task and ensure that the data they provide can be processed efficiently by the robot's control systems.

7.7. Mobile and Portable Power Solutions

In our increasingly mobile world, the demand for portable power solutions has never been greater. From smartphones to laptops, portable electronic devices rely on efficient and reliable power sources to keep us connected, productive, and entertained on the go. Electrical engineering plays a crucial role in developing these systems, particularly through advances in battery technology and renewable energy integration. In this section, we will explore the fundamentals of mobile power solutions, with a focus on batteries and solar chargers, to understand how engineers solve the challenges of energy storage and distribution.

The Importance of Portable Power Solutions

Portable power systems are essential in today's digital age. Whether you're traveling, working remotely, or spending time outdoors, having a reliable power source for your devices is critical. Electrical engineers design and optimize these power solutions to be lightweight, efficient, and long-lasting, ensuring that devices can operate wherever and whenever needed. These systems encompass multiple technologies:

- **Batteries**: Batteries are the most common portable power source, used in everything from smartphones to electric vehicles. Engineers continually work to improve battery life, capacity, and safety, focusing on technologies like lithium-ion and solid-state batteries (as introduced in Chapter 5, where we discussed the basics of energy storage).

- **Solar Chargers**: Solar chargers provide a renewable way to power devices, converting sunlight into electricity. These are particularly useful for outdoor activities or in regions with limited access to traditional power sources.

- **Power Banks**: Power banks are portable battery packs that store energy and can recharge devices on the go. They are designed to be compact and efficient, providing a convenient backup power source, which reflects the growing demand for autonomy in energy management, an essential topic discussed in earlier chapters.

Key Components of Mobile Power Systems

Mobile power systems rely on several key components working together to provide portable, reliable energy:

- **Rechargeable Batteries**: At the core of any portable power solution, rechargeable batteries store energy and deliver it to devices as needed. Lithium-ion batteries are the most common, due to their high energy density and rechargeability.

- **Charge Controllers**: These devices regulate the flow of electricity into and out of the battery, protecting it from overcharging and ensuring efficient energy use. Charge controllers are essential in systems like solar chargers, where power input from the sun can vary throughout the day.

- **Inverters and DC-DC Converters**: Inverters convert direct current (DC) from batteries into alternating current (AC), which is used by most household appliances. DC-DC converters, on the other hand, adjust the voltage of DC power to match the needs of specific devices, optimizing the energy flow.

- **Solar Panels**: Solar panels capture sunlight and convert it into electricity through photovoltaic cells. The size, placement, and efficiency of the panels play a crucial role in determining how much energy they can generate and how quickly they can recharge a battery. A deeper understanding of how solar energy is harnessed and its practical applications were discussed earlier in this chapter when we explored renewable energy systems.

Learning How Solar Charging Works

Let's explore the theoretical foundations of a typical solar charger system and understand how electrical engineers design such systems to maximize energy efficiency.

System Overview: A solar charger system consists of several interrelated components:

1. **Solar Panel Setup**: Solar panels are placed in areas where they can capture maximum sunlight. Electrical engineers must consider variables like panel orientation, shading, and seasonal variations in sunlight exposure.

2. **Charge Controllers**: These controllers regulate voltage and current from the solar panels, ensuring that the battery is charged safely and efficiently. A key design consideration is how to prevent overcharging and optimize the panel's output even under varying light conditions.

3. **Battery Storage**: Batteries store the energy produced by the solar panels, allowing devices to be powered when sunlight isn't available. Engineers must calculate the appropriate battery capacity based on expected usage patterns, device power requirements, and solar panel output.

4. **USB Output and Voltage Regulation**: The stored energy is converted to the correct voltage for charging mobile devices, usually via a USB output. Engineers design circuits to step down or regulate voltage for safe and efficient charging.

This learning exercise on solar chargers provides insights into how modern portable energy solutions are designed, and how different components work together to manage power from renewable sources.

How Portable Power Systems Work

Here's a simplified breakdown of how solar-based portable power systems operate:

- **Energy Generation**: The solar panel captures energy from sunlight, converting it into DC electricity.

- **Energy Regulation**: The charge controller regulates the voltage and current coming from the solar panel to prevent damage to the battery.

- **Energy Storage**: The rechargeable battery stores the regulated energy for use when direct sunlight is not available.

- **Energy Use**: A voltage regulation system ensures that the battery provides the appropriate voltage to charge devices like smartphones, laptops, or other portable gadgets.

Practical Knowledge: Applications in Real-World Portable Power Solutions

Understanding how portable power systems work in theory has numerous practical applications:

- **Portable Solar Generators**: Larger solar generators apply similar technology on a bigger scale, providing power for camping trips, emergency situations, or off-grid living. They can charge multiple devices or even power small appliances, integrating multiple solar panels and higher-capacity batteries.

- **Wearable Solar Technology**: Solar panels integrated into clothing or backpacks provide a convenient way to charge devices while on the move, representing the fusion of electrical engineering and modern consumer lifestyle needs.

- **Hybrid Systems**: Combining solar power with traditional battery packs creates hybrid systems, extending power availability for remote or long-duration activities. These systems are critical for applications in areas without reliable grid power, emphasizing the importance of energy autonomy in today's world.

7.8. Curiosities and Further Insights

The field of electrical engineering is full of fascinating stories, advanced applications, and emerging technologies that continue to shape the modern world. This section explores some interesting facts and insights related to the concepts we've covered in this chapter, offering a glimpse into the broader impact and future potential of electrical engineering.

The Origins of Wireless Communication

Wireless communication, a cornerstone of modern technology, began with the work of pioneers like Nikola Tesla and Guglielmo Marconi. In the late 19th and early 20th centuries, their experiments with radio waves laid the groundwork for the development of wireless telegraphy, radio, and eventually, all forms of wireless communication we rely on today, from smartphones to Wi-Fi.

Interesting Fact: Marconi's first successful transatlantic radio communication in 1901 used equipment that operated at a frequency of about 820 kHz, far below the frequencies used by modern wireless devices, which can operate in the gigahertz range.

The Rise of Smart Grids

As the demand for electricity grows and the need for sustainable energy becomes more pressing, the concept of the smart grid has emerged. A smart grid uses digital communication technology to detect and react to local changes in usage, improving the efficiency and reliability of electricity distribution. Smart grids are also essential for integrating renewable energy sources, such as solar and wind power, into the existing infrastructure.

Interesting Fact: One of the first large-scale implementations of smart grid technology was in Italy in the early 2000s. The country installed over 30 million smart meters, making it one of the most advanced electricity networks in the world at the time.

The Impact of Moore's Law on Consumer Electronics

Moore's Law, named after Intel co-founder Gordon Moore, predicts that the number of transistors on a microchip will double approximately every two years, leading to exponential growth in computing power. This trend has driven the rapid advancement of consumer electronics, enabling the miniaturization of devices and the proliferation of high-performance smartphones, tablets, and laptops.

Interesting Fact: The iPhone 12, released in 2020, contains a chip with nearly 12 billion transistors, compared to the first Intel processor in 1971, which had just 2,300 transistors.

Renewable Energy's Rapid Growth

The shift toward renewable energy has accelerated in recent years, with solar and wind power leading the way. Advances in technology, driven by electrical engineering, have made renewable energy sources more efficient and cost-effective, contributing to their widespread adoption.

Interesting Fact: As of 2020, solar energy was the cheapest source of electricity in history, according to the International Energy Agency (IEA). The cost of solar power has fallen by nearly 90% over the past decade, making it a key player in the global energy transition.

The Future of Robotics in Everyday Life

Robotics is no longer confined to industrial settings; it is increasingly becoming part of everyday life. From robotic vacuum cleaners to drones and automated delivery systems, robots are taking on a variety of tasks that were once the domain of humans. The development of AI and machine learning is further enhancing the capabilities of robots, making them more autonomous and adaptable.

Interesting Fact: In Japan, where the aging population is creating a demand for elder care, robots are being developed to assist with daily tasks such as lifting patients, providing companionship, and even administering medication.

The Evolution of Battery Technology

Battery technology is evolving rapidly, driven by the demand for longer-lasting, faster-charging, and more efficient energy storage solutions. Solid-state batteries, which use a solid electrolyte instead of a liquid one, are one of the most promising developments. They offer higher energy density, faster charging times, and improved safety compared to traditional lithium-ion batteries.

Interesting Fact: The first generation of electric vehicles (EVs) had a range of around 100 miles per charge. Thanks to advances in battery technology, modern EVs can now travel over 300 miles on a single charge, with some models exceeding 400 miles.

The Internet of Things (IoT) and Its Expanding Role

The Internet of Things (IoT) is revolutionizing industries by connecting everyday objects to the internet, enabling them to send and receive data. This connectivity allows for greater automation, efficiency, and data-driven decision-making in sectors ranging from agriculture to healthcare.

Interesting Fact: By 2025, it is estimated that there will be over 75 billion IoT devices worldwide, ranging from smart home appliances to industrial sensors, transforming how we interact with the world around us.

Chapter 8

Safety and Maintenance

8.1. Principles of Electrical Safety

Electrical safety is a critical aspect of engineering practice. Whether working on a small household project or managing large industrial systems, understanding and applying the principles of electrical safety is essential to prevent accidents, injuries, and even fatalities. This section covers the fundamental concepts of electrical safety, including the recognition of electrical hazards, adherence to safety standards and regulations, the proper use of personal protective equipment (PPE), and the implementation of safe work practices.

Understanding Electrical Hazards

Electrical hazards are potential dangers that can arise from the improper use, installation, or maintenance of electrical systems. These hazards can cause serious harm, including electric shocks, burns, fires, and explosions. Understanding the types of electrical hazards and how they occur is the first step in preventing accidents.

- **Electric Shock:** Occurs when a person comes into contact with a live electrical conductor, causing current to flow through their body. The severity of the shock depends on the current's magnitude, the path it takes through the body, and the duration of exposure. Proper grounding (discussed in detail in Chapter 2: Grounding and Earthing Systems) is one of the most effective ways to prevent electric shocks by providing a safe path for electrical current to flow to the Earth, reducing the risk of dangerous current passing through a person.

- **Arc Flash**: A sudden release of energy caused by an electrical fault. It can result in intense heat, light, and pressure, leading to severe burns, blindness, and even death.

- **Electrical Burns**: Result from the heat generated by electrical current as it passes through the body or other materials. These burns can occur both at the point of contact and internally, where the current travels through tissue.

- **Fire and Explosion**: Electrical equipment that is improperly maintained, overloaded, or damaged can ignite flammable materials, leading to fires or explosions. These incidents can be catastrophic, especially in industrial settings.

Safety Standards and Regulations

Compliance with safety standards and regulations is essential in all electrical work. These standards are designed to minimize risk and ensure that electrical systems are safe to operate. Different countries have specific regulations, but many align with international standards.

- **National Electrical Code (NEC)**: In the United States, the NEC sets the standard for safe electrical installation and is adopted by most states. It covers the design, installation, and inspection of electrical systems to prevent hazards.

- **Occupational Safety and Health Administration (OSHA)**: OSHA enforces workplace safety regulations, including those related to electrical hazards. OSHA standards require proper training, use of PPE, and adherence to safe work practices.

- **International Electrotechnical Commission (IEC)**: The IEC develops global standards for electrical technologies. These standards are used in many countries to ensure consistency and safety in electrical engineering practices.

- **European Committee for Electrotechnical Standardization (CENELEC)**: CENELEC is responsible for European standards, ensuring safety and compatibility across electrical systems in Europe.

Personal Protective Equipment (PPE)

Personal Protective Equipment (PPE) is essential for protecting workers from electrical hazards. The correct selection and use of PPE can significantly reduce the risk of injury in electrical work.

- **Insulated Gloves**: Protect hands from electric shocks and burns. Insulated gloves are rated for different voltage levels, and it's crucial to select the appropriate type for the job.

- **Arc Flash Clothing**: Specialized clothing that protects against the intense heat of an arc flash. This includes flame-resistant (FR) shirts, pants, and face shields designed to withstand high temperatures.

- **Safety Helmets**: Protect the head from falling objects and potential electrical hazards. Helmets with arc-rated face shields provide additional protection against arc flash injuries.

- **Safety Glasses and Face Shields**: Protect the eyes and face from flying debris, arc flashes, and other hazards. They should be used whenever working near live electrical equipment.

- **Insulated Tools**: Tools with insulated handles reduce the risk of electric shock when working with live circuits. Only tools that are specifically rated for electrical work should be used.

Safe Work Practices in Electrical Engineering

Adopting safe work practices is crucial for minimizing risk and ensuring a safe working environment. These practices involve proper planning, use of tools, and adherence to procedures designed to protect workers and equipment.

- **Lockout/Tagout (LOTO)**: A safety procedure that ensures machines and circuits are properly shut off and cannot be turned on again before maintenance or repair work is completed. LOTO involves locking the equipment in the off position and tagging it with a warning label.

- **De-energization**: Whenever possible, work on de-energized circuits to eliminate the risk of electric shock. This involves disconnecting the power supply and verifying that the circuit is not live before starting work.

- **Use of Ground Fault Circuit Interrupters (GFCIs)**: GFCIs are devices that detect ground faults and interrupt the flow of electricity, preventing shocks. They are particularly important in wet or outdoor environments.

- **Proper Tool Use and Maintenance**: Use tools that are specifically designed for electrical work and ensure they are in good condition. Regular inspection and maintenance of tools are essential to prevent accidents.

- **Awareness and Training**: Continuous education and training are vital for staying updated on safety practices and regulations. Workers should be trained to recognize hazards, use PPE correctly, and follow safe work procedures.

8.2. Basic Maintenance of Electrical Systems

Maintenance is a critical aspect of electrical engineering, ensuring that systems operate efficiently, safely, and reliably over time. Regular maintenance helps prevent unexpected failures, extend the lifespan of equipment, and ensure compliance with safety regulations. This section covers the essential aspects of electrical system maintenance, including routine inspections, troubleshooting common issues, implementing preventive maintenance strategies, and the importance of proper documentation and record-keeping.

Routine Inspection and Maintenance

Routine inspection and maintenance are essential for identifying potential issues before they lead to failures. Regular checks ensure that all components of an electrical system are functioning correctly and safely.

- **Visual Inspections**: Regular visual inspections help detect obvious signs of wear and tear, such as frayed wires, loose connections, corrosion, or damaged insulation. These inspections should be performed frequently to catch issues early.

- **Thermal Imaging**: Using thermal imaging cameras during inspections can help identify hotspots that indicate overheating components or loose connections. These issues can be addressed before they lead to more significant problems, such as equipment failure or fire.

- **Testing and Measurements**: Regularly testing electrical systems with appropriate instruments (e.g., multimeters, insulation resistance testers) ensures that circuits and components are operating within their specified parameters. Measurements should include checking voltage, current, resistance, and insulation quality.

- **Cleaning and Lubrication**: Dust, dirt, and moisture can accumulate on electrical components, leading to overheating or short circuits. Regular cleaning of electrical panels, connectors, and components is essential. Lubrication of moving parts, such as switches or motor bearings, can also prevent mechanical failure.

- **Battery Maintenance**: For systems that include batteries (e.g., uninterruptible power supplies or solar systems), regular maintenance involves checking charge levels, cleaning terminals, and ensuring proper ventilation. Batteries should be replaced according to the manufacturer's recommendations to prevent failures.

Troubleshooting Common Electrical Problems

Even with regular maintenance, electrical systems can encounter issues that require troubleshooting. Understanding common problems and their solutions is key to maintaining system reliability.

- **Short Circuits**: A short circuit occurs when a live wire touches a neutral or ground wire, causing excessive current flow. This can trip circuit breakers or cause equipment damage. Troubleshooting involves locating the fault, repairing the damaged wiring, and resetting the circuit breaker.

- **Overloaded Circuits**: Circuits that carry more current than they are designed for can overheat, potentially leading to fires. Overloads are often caused by too many devices plugged into a single circuit. The solution involves redistributing the load across multiple circuits or upgrading the wiring to handle the higher current.

- **Ground Faults**: Ground faults occur when electrical current takes an unintended path to the ground, often through a person's body. Ground Fault Circuit Interrupters (GFCIs) can detect and interrupt ground faults. Troubleshooting involves locating and repairing the fault in the wiring or equipment.

- **Voltage Fluctuations**: Inconsistent voltage can cause equipment to malfunction or get damaged. Voltage fluctuations can be due to loose connections, faulty transformers, or issues with the power supply. Identifying and correcting the source of the fluctuation is essential to maintaining stable operation.

- **Nuisance Tripping**: Circuit breakers or GFCIs that trip without an apparent cause may indicate underlying issues such as deteriorating insulation or faulty wiring. Troubleshooting involves inspecting the affected circuit and testing components to identify the root cause.

Preventive Maintenance Strategies

Preventive maintenance involves regular, planned activities designed to prevent equipment failure and extend the life of electrical systems. This proactive approach is more cost-effective than reactive maintenance, which only addresses issues after they occur.

- **Scheduled Maintenance**: Establish a regular schedule for inspecting and maintaining electrical systems. This schedule should be based on the manufacturer's recommendations, the operating environment, and the criticality of the equipment.

- **Component Replacement**: Replace components such as circuit breakers, fuses, and relays before they reach the end of their expected

Safety and Maintenance

life. Proactive replacement prevents unexpected failures and reduces downtime.

- **System Upgrades**: As technology advances, upgrading older systems with more efficient or safer components can improve performance and reduce maintenance needs. For example, replacing outdated wiring or installing more efficient lighting can reduce energy consumption and improve safety.

- **Training and Awareness**: Ensure that all personnel involved in maintenance are properly trained and aware of the latest safety protocols and procedures. Regular training sessions can prevent accidents and ensure that maintenance tasks are performed correctly.

Documentation and Record-Keeping

Proper documentation and record-keeping are crucial components of effective maintenance management. Keeping accurate records ensures that maintenance is performed regularly and provides a history that can be invaluable in diagnosing future issues.

- **Maintenance Logs**: Maintain a log of all maintenance activities, including inspections, repairs, and component replacements. This log should include the date, details of the work performed, and the names of the personnel involved.

- **Equipment History**: Keep detailed records of each piece of equipment, including installation dates, maintenance history, and any modifications or upgrades. This information is critical for planning future maintenance and making informed decisions about repairs or replacements.

- **Compliance Records**: Maintain records of compliance with safety standards and regulations, including any inspections or audits performed by regulatory agencies. These records are essential for demonstrating adherence to legal requirements and for passing inspections.

- **Software Tools**: Consider using maintenance management software to track and schedule maintenance activities. These tools can automate reminders, generate reports, and help manage large amounts of data efficiently.

8.3. Guidelines for Home and Workplace Safety

Electrical safety is a critical concern in both home and workplace environments. Understanding and following established safety guidelines can prevent accidents, injuries, and damage to property. This section provides practical advice on maintaining electrical safety in these settings, covering common hazards, essential safety procedures, and how to respond effectively in emergencies.

Electrical Safety in the Home

Many electrical hazards in the home can be prevented with proper awareness and maintenance. Homeowners should be proactive in identifying potential risks and taking steps to mitigate them.

- **Inspect Electrical Outlets**: Regularly check electrical outlets for signs of wear, such as cracks, discoloration, or loose connections. Replace any damaged outlets to prevent short circuits or fires.

- **Avoid Overloading Circuits**: Plugging too many devices into a single outlet or power strip can overload the circuit, leading to overheating and potential fires. Spread the load across multiple outlets and use surge protectors to manage power distribution safely.

- **Childproof Outlets**: Use outlet covers or tamper-resistant outlets to prevent children from inserting objects into electrical outlets. This simple step can significantly reduce the risk of electric shock for young children.

- **Use GFCIs in Wet Areas**: Ground Fault Circuit Interrupters (GFCIs) should be installed in areas where water is present, such as bathrooms, kitchens, and outdoor outlets. GFCIs quickly shut off power if a ground fault is detected, preventing electric shocks.

- **Unplug Appliances When Not in Use**: Unplugging appliances when they are not in use reduces the risk of electrical fires. This is particularly important for high-wattage devices like space heaters and irons.

Safety and Maintenance

- **Keep Electrical Appliances Away from Water**: Water is a conductor of electricity, and using electrical appliances near water increases the risk of shock. Always keep appliances like hairdryers, radios, and phone chargers away from sinks, bathtubs, and other water sources.

Workplace Safety Procedures

In the workplace, electrical safety is governed by strict regulations and requires diligent attention to detail. Employers and employees must work together to ensure a safe working environment.

- **Conduct Regular Safety Audits**: Regularly scheduled safety audits help identify potential electrical hazards before they lead to accidents. These audits should be comprehensive, covering all electrical systems, equipment, and work practices.

- **Follow Lockout/Tagout (LOTO) Procedures**: When performing maintenance or repairs on electrical equipment, always follow LOTO procedures. This ensures that equipment is properly de-energized and cannot be accidentally re-energized while work is being performed.

- **Use Proper Signage**: Clearly mark high-voltage areas, electrical panels, and other hazardous locations with appropriate warning signs. Signage helps prevent accidental contact with dangerous equipment.

- **Maintain Clear Access to Electrical Panels**: Ensure that electrical panels and circuit breakers are easily accessible and not blocked by equipment, furniture, or other obstructions. In an emergency, quick access to these panels is essential for shutting off power.

- **Wear Appropriate PPE**: Personal Protective Equipment (PPE) is essential for protecting workers from electrical hazards. This includes insulated gloves, safety glasses, flame-resistant clothing, and arc flash protection. Always wear the correct PPE for the task at hand.

- **Employee Training and Education**: Regular training sessions ensure that employees are aware of the latest safety practices and understand how to recognize and respond to electrical hazards. Training should cover the safe use of equipment, emergency procedures, and the importance of PPE.

Emergency Response for Electrical Accidents

Despite the best precautions, electrical accidents can still occur. Knowing how to respond quickly and effectively can minimize the severity of injuries and prevent further harm.

- **Electrical Shock**: If someone receives an electric shock, do not touch them directly if they are still in contact with the electrical source. First, disconnect the power by shutting off the circuit breaker or unplugging the device. Once the person is no longer in contact with the source, call emergency services immediately and provide first aid, including CPR if necessary.

- **Electrical Fire**: In the event of an electrical fire, never use water to extinguish the flames, as it can conduct electricity and cause further harm. Instead, use a Class C fire extinguisher, which is designed for electrical fires. If the fire is small and contained, you can attempt to extinguish it. Otherwise, evacuate the area and call the fire department.

- **Burns and Injuries**: Treat electrical burns by cooling the affected area with cool, running water for at least 10 minutes. Do not apply ice or ointments, as these can worsen the injury. Cover the burn with a clean, dry cloth and seek medical attention immediately.

- **Power Outages**: During a power outage, turn off and unplug non-essential appliances to prevent damage from power surges when electricity is restored. Use flashlights instead of candles to reduce the risk of fire, and avoid opening refrigerators and freezers to keep food cold for as long as possible.

Safety Audits and Compliance

Regular safety audits are crucial for maintaining a safe environment in both home and workplace settings. These audits help ensure compliance with safety regulations and identify areas where improvements can be made.

- **Develop a Safety Audit Checklist**: Create a comprehensive checklist that covers all aspects of electrical safety, including equipment inspection, PPE usage, emergency preparedness, and compliance with local regulations.

- **Schedule Regular Audits**: Conduct audits on a regular basis, at least annually, or more frequently in high-risk environments. Regular audits help ensure that safety practices are being followed consistently.

- **Document Findings and Actions**: Record the findings of each audit, including any safety violations or areas for improvement. Document the actions taken to address these issues and ensure that follow-up audits verify the effectiveness of these actions.

- **Stay Informed of Regulatory Changes**: Safety regulations can change over time, and it's important to stay informed of any updates that may affect your safety practices. Regularly review and update your safety protocols to remain in compliance with the latest standards.

Electrical safety is a shared responsibility that requires vigilance, education, and adherence to best practices. By following these guidelines for home and workplace safety, you can reduce the risk of accidents and create a safer environment for everyone. Regular audits, proper training, and the use of appropriate protective equipment are key components of an effective safety program.

Access Your High-Resolution PDF Version

Optimized Formatting and Easy Reference

Thank you for purchasing this book!

I understand that many readers prefer to study and take notes on different devices, but the formatting of printed and digital books can vary, affecting the readability of images, diagrams, and formulas.

That's why I offer the PDF version for free to those who purchase the book.

This allows you to access it anywhere, quickly search for specific concepts, add annotations, and benefit from a consistently clear and optimized layout for study and work.

To get your free PDF version, simply send an email with the subject:

"Electrical Engineering PDF Version Request" to:

gregorhaynesauthor@gmail.com

Please attach a screenshot of your purchase confirmation or proof of order from Amazon. Once received, I'll promptly send you the high-quality **PDF version of my book**, ensuring you have the best reading experience.

If you found this book valuable, I would truly appreciate your honest review on Amazon.

Your feedback not only helps other readers make informed decisions but also supports the continuous improvement of this book.

Your insights are invaluable, and every review makes a difference!

Appendices

Glossary of Technical Terms

This glossary provides definitions for key terms and concepts used throughout the book. Understanding these terms is essential for mastering the principles of electrical engineering.

Alternating Current (AC): A type of electrical current in which the direction of flow reverses periodically. AC is commonly used for power distribution in homes and industries.

Ammeter: A device used to measure the electric current in a circuit, connected in series with the load.

Ampere (A): The unit of electric current in the International System of Units (SI). One ampere represents the flow of one coulomb of charge per second.

Arc Flash: A dangerous condition associated with the release of energy caused by an electric arc. It can result in severe burns, injuries, and damage to equipment.

Battery: A device consisting of one or more electrochemical cells that store and supply electrical energy. Batteries are used to power a wide range of electronic devices.

Capacitance: The ability of a system to store electrical charge, measured in farads.

Capacitor: An electrical component that stores energy in an electric field, used to smooth voltage fluctuations and store energy temporarily in circuits.

Circuit Breaker: An automatic electrical switch designed to protect an electrical circuit from damage caused by overcurrent or short circuits. It interrupts the current flow upon detecting a fault.

Conductance: The measure of how easily electricity flows through a material. It is the inverse of resistance and is measured in siemens (S).

Conductor: A material that permits the flow of electrical current, commonly used in wiring and electrical components.

Appendices

Coulomb (C) – The unit of electric charge in the International System of Units (SI), representing the quantity of charge transferred by a current of one ampere in one second.

Current: The flow of electric charge in a conductor, typically measured in amperes (A). Current can be either direct (DC) or alternating (AC).

Direct Current (DC): A type of electrical current in which the flow of charge is unidirectional, meaning it flows in one direction only. DC is commonly used in batteries and electronic devices.

Diode: A semiconductor device that allows current to flow in one direction only. Diodes are used in rectifiers, voltage regulation, and other applications.

Electromagnet: A magnet created by passing an electric current through a coil of wire, used in various devices like motors and relays.

Electromagnetic Field: A field produced by moving electric charges that can affect the behavior of other charged particles.

Electromagnetism: The study of the interaction between electric currents and magnetic fields, one of the fundamental forces of nature.

Electromotive Force (EMF): The voltage generated by a source like a battery or generator, driving the flow of electric current.

Energy Efficiency: The ratio of useful energy output to the total energy input, expressed as a percentage.

Farad (F): The unit of capacitance in the International System of Units (SI). One farad represents the capacitance that stores one coulomb of charge per volt of potential difference.

Frequency (Hz): The number of cycles per second in an alternating current, measured in hertz.

Fuse: A safety device that protects electrical circuits by melting and breaking the circuit when the current exceeds a safe level.

Ground (Earth): A reference point in an electrical circuit, used to safely discharge excess electricity.

Ground Fault: An unintentional connection between a live electrical conductor and the ground, often resulting in electric shock.

Ground Fault Circuit Interrupter (GFCI): A device designed to protect people from electric shock by detecting ground faults and interrupting the flow of electricity.

Henry (H): The unit of inductance in the International System of Units (SI). It represents the amount of inductance in a circuit where a change in current of one ampere per second induces a voltage of one volt.

Hertz (Hz): The unit of frequency in the International System of Units (SI), representing the number of cycles per second of a periodic signal or waveform. One hertz equals one cycle per second.

HVAC Systems: Heating, Ventilation, and Air Conditioning systems that control indoor environmental conditions.

IEC (International Electrotechnical Commission): A global organization that develops international standards for electrical and electronic technologies.

Impedance (Z): The total opposition to the flow of alternating current in a circuit, consisting of both resistance and reactance. It is measured in ohms (Ω).

Inductance (H): The property of a conductor by which a change in current through it induces an electromotive force (EMF) in both the conductor itself and any nearby conductors.

Inductor: A component in electrical circuits that stores energy in a magnetic field when current flows through it.

Insulator: A material that resists the flow of electric current, used to protect against electric shocks and to prevent short circuits.

Integrated Circuit (IC): A set of electronic circuits built on a small semiconductor chip, used in nearly all modern electronic devices.

Joule (J): The unit of energy in the International System of Units (SI). One joule is equal to the energy transferred when a force of one newton is applied over a distance of one meter.

Kilowatt-hour (kWh): A unit of energy commonly used for measuring electrical energy consumption. One kilowatt-hour is equivalent to using 1,000 watts of power for one hour.

Kirchhoff's Laws: Two fundamental laws in electrical engineering: Kirchhoff's Current Law (KCL) states that the total current entering a junction equals the total current leaving

Appendices

it. Kirchhoff's Voltage Law (KVL) states that the sum of all voltages around a closed loop equals zero.

LED (Light-Emitting Diode): A semiconductor device that emits light when an electric current passes through it, used in displays and lighting.

Load: The component or device in a circuit that consumes electrical power.

Magnetic Flux: A measure of the amount of magnetic field passing through a given area, used to describe the strength and direction of a magnetic field.

Magnetic Resonance Imaging (MRI) Machines: Medical imaging devices that use strong magnetic fields and radio waves to produce detailed images of internal body structures.

MOSFET (Metal-Oxide-Semiconductor Field-Effect Transistor): A type of transistor used for amplifying or switching electronic signals, widely used in digital and analog circuits.

Multimeter: A versatile measuring instrument that can measure voltage, current, and resistance in electrical circuits.

Ohm (Ω): The unit of electrical resistance in the International System of Units (SI). One ohm is the resistance between two points in a conductor when a constant potential difference of one volt produces a current of one ampere.

Ohm's Law: A fundamental principle in electrical engineering that states the relationship between voltage (V), current (I), and resistance (R): $V = IR$

Oscillator: A circuit that generates a continuous, repetitive electronic signal, often used in clocks, radios, and computers.

Oscilloscope: An instrument used to visualize electrical signals over time, typically used for observing the waveform of AC signals and diagnosing circuit behavior.

Photodiode: A semiconductor device that converts light into electrical current, commonly used in sensors and light detection systems.

Photovoltaic (PV) Cell: A semiconductor device that converts light energy directly into electrical energy, commonly used in solar panels.

PN Junction: The boundary between P-type and N-type semiconductor materials, essential in devices such as diodes and transistors.

Power (W): The rate at which electrical energy is transferred, measured in watts.

Power Factor: The ratio of real power used in a circuit to the apparent power supplied to the circuit, a measure of how effectively electrical power is being used.

Rectifier: A device that converts alternating current (AC) to direct current (DC), commonly used in power supplies.

Relay: An electrically operated switch used to control a circuit by a separate low-power signal, often used for switching high currents or voltages.

Resistance: The opposition to the flow of electric current in a circuit, measured in ohms.

Resistor: An electrical component that opposes the flow of current, used to control voltage and current in a circuit.

Schottky Diode: A diode with a low forward voltage drop, commonly used for high-speed switching applications.

Semiconductor: A material with electrical conductivity between that of a conductor and an insulator, used in devices like diodes, transistors, and integrated circuits.

Short Circuit: A fault in an electrical circuit where a low-resistance connection is made between two points, leading to excessive current flow and potential damage.

Signal Ground: A reference point in a circuit that provides a common return path for electrical signals to prevent interference.

Sine Wave: A smooth periodic oscillation that describes alternating current (AC) and other waveforms in electrical engineering.

Solar Panels: Devices made up of photovoltaic cells that convert sunlight into electrical energy.

Solid-State Relays (SSRs): Relays that use semiconductor devices to switch electrical loads without moving parts, allowing faster switching and longer life.

Surface Mount Technology (SMT): A method for mounting electronic components directly onto the surface of a printed circuit board.

Switch: A device that opens or closes a circuit, allowing or preventing the flow of electrical current.

Temperature Coefficient: The rate at which a material's resistance changes with temperature.

Thermistor: A type of resistor whose resistance varies significantly with temperature, used in temperature sensing and protection.

Transformer: A device that transfers electrical energy between two or more circuits through electromagnetic induction, commonly used to step up or step down voltage levels.

Transistor: A semiconductor device used to amplify or switch electronic signals, a fundamental building block of modern electronics.

Voltage (V): The electric potential difference between two points, which drives the flow of current in a circuit. It is measured in volts (V).

Watt (W): The unit of power in the International System of Units (SI). One watt is equal to one joule of energy per second.

Wire: A conductor, usually made of copper or aluminum, used to carry electrical current.

Zener Diode: A type of diode designed to allow current to flow in the reverse direction when a specific reverse voltage is reached, used in voltage regulation.

Reference Tables and Constants

This section provides essential reference tables and constants that are frequently used in electrical engineering. These tables will be useful for calculations, designing circuits, and understanding the properties of electrical components.

SI Units and Prefixes

Prefix	Symbol	Factor
Tera	T	10^{12}
Giga	G	10^{9}
Mega	M	10^{6}
Kilo	k	10^{3}
Hecto	h	10^{2}
Deca	da	10^{1}
(Base)		10^{0}
Deci	d	10^{-1}
Centi	c	10^{-2}
Milli	m	10^{-3}
Micro	μ	10^{-6}
Nano	n	10^{-9}
Pico	p	10^{-12}

Appendices

Common Electrical Constants

Constant	Symbol	Value	Unit
Speed of Light in Vacuum	c	3.00×10^{8}	m/s
Elementary Charge	e	1.602×10^{-19}	C (Coulombs)
Permittivity of Free Space	ε_0	8.854×10^{-12}	F/m
Permeability of Free Space	μ_0	$4\pi \times 10^{-7}$	H/m
Planck's Constant	h	6.626×10^{-34}	J·s
Boltzmann Constant	k	1.381×10^{-23}	J/K
Avogadro's Number	N_a	6.022×10^{23}	mol^{-1}
Universal Gas Constant	R	8.314	J/(mol·K)
Gravitational Constant	G	6.674×10^{-11}	N·m²/kg²
Electron Rest Mass	m_e	9.109×10^{-31}	kg
Proton Rest Mass	m_p	1.673×10^{-27}	kg
Neutron Rest Mass	m_n	1.675×10^{-27}	kg
Stefan-Boltzmann Constant	σ	5.670×10^{-8}	W/(m²·K⁴)

Resistivity of Common Materials

Material	Resistivity (ϱ)	Unit
Silver	1.59×10^{-8}	$\Omega \cdot m$
Copper	1.68×10^{-8}	$\Omega \cdot m$
Gold	2.44×10^{-8}	$\Omega \cdot m$
Aluminum	2.82×10^{-8}	$\Omega \cdot m$
Tungsten	5.60×10^{-8}	$\Omega \cdot m$
Iron	9.71×10^{-8}	$\Omega \cdot m$
Platinum	10.6×10^{-8}	$\Omega \cdot m$
Nichrome	1.10×10^{-6}	$\Omega \cdot m$
Carbon (Graphite)	3.50×10^{-5}	$\Omega \cdot m$
Silicon (Intrinsic)	6.40×10^{2}	$\Omega \cdot m$

Capacitor Color Code (Standard EIA Code)

Color	Digit	Multiplier	Tolerance (%)
Black	0	×1	
Brown	1	×10	±1
Red	2	×100	±2
Orange	3	×1,000	
Yellow	4	×10,000	
Green	5	×100,000	±0.5
Blue	6	×1,000,000	±0.25
Violet	7	×10,000,000	±0.1
Gray	8	×100,000,000	
White	9	×1,000,000,000	
Gold		×0.1	±5
Silver		×0.01	±10
None			±20

Power Ratings of Resistors

Resistor Power Rating	Dimensions (L×D)	Typical Application
1/8 Watt	3.2mm × 1.6mm	Low-power signal processing circuits
1/4 Watt	6.0mm × 2.3mm	General-purpose applications
1/2 Watt	9.0mm × 3.2mm	Power supply circuits
1 Watt	11.0mm × 4.5mm	High-power circuits
2 Watts	15.0mm × 5.0mm	High-current or heat-sensitive areas

Standard Wire Gauges (AWG)

AWG	Diameter (mm)	Cross-Sectional Area (mm^2)	Resistance per Meter (Ω)
10	2.588	5.261	0.00328
12	2.053	3.309	0.00521
14	1.628	2.081	0.00829
16	1.291	1.308	0.0132
18	1.024	0.823	0.0209
20	0.812	0.518	0.0332
22	0.644	0.326	0.0524
24	0.511	0.205	0.0836
26	0.404	0.129	0.133

Electronic Circuit Symbols

WIRE			TRANSISTORS			DIODES	
WIRE			NPN BJT			DIODE	
JOINED WIRE			PNP BJT			LED	
			PNP JFET			ZENER DIODE	
WIRE NOT JOINED			NPN JFET			PHOTO DIODE	
			N CHANNEL MOSFET enh			SCR	
			P CHANNEL MOSFET enh			VARICAP	
POWER SUPPLY	CELL		N CHANNEL MOSFET enh			TUNNEL DIODE	
	BATTERY		P CHANNEL MOSFET enh			SCHOTTKY DIODE	
			N CHANNEL MOSFET dep		CAPACITORS	CAPACITOR	
	DC SUPPLY		P CHANNEL MOSFET dep			CAPACITOR POLARISED	
	AC SUPPLY		PHOTO-TRANSISTOR			VARIABLE CAPACITOR	
	EARTH/ GROUND		METERS	VOLTMETER		TRIMMER CAPACITOR	
				AMMETER	RESISTORS	RESISTOR	
	FUSE			GALVANOMETER		RHEOSTAT	
				OHMMETER		POTENTIOMETER	
				OSCILLOSCOPE		PRESET	

Appendices

159

Appendices

SWITCHES	PUSH TO BREAK		**INPUT DEVICES**	MICROPHONE		**LOGIC GATES**	NOT
	PUSH TO MAKE			THERMISTOR			
	SPST			LDR			AND
	SPDT						
	SPTT			PHONE JACK			NAND
	DPST						
	DPDT			TRANSFORMER			OR
				TRANSFORMER WITH CENTRE TAP			
OUTPUT DEVICES	BELL		**MISCELLANEOUS**	AMPLIFIER			NOR
	BUZZER			ANTENAE			
	EARPHONE			RELAY			EX-OR
	SPEAKER						
	LAMP			INTEGRATED CIRCUIT			EX-NOR
	LAMP						
	MOTOR						
	HEATER						
	INDUCTOR						
	PIEZO TRANSDUCER						

Appendices

Resistor Color Code (5 Bands)

Color	1st Digit	2nd Digit	3rd Digit	Multiplier	Tolerance
Black	0	0	0	x 1	
Brown	1	1	1	x 10	1%
Red	2	2	2	x 100	2%
Orange	3	3	3	x 1,000	3%
Yellow	4	4	4	x 10,000	4%
Green	5	5	5	x 100,000	0.5%
Blue	6	6	6	x 1,000,000	0.25%
Violet	7	7	7	x 10,000,000	0.10%
Grey	8	8	8	x 100,000,000	0.05%
White	9	9	9	x 1,000,000,000	
Gold				x 0,1	5%
Silver				x 0,01	10%
None					20%

Index

absolute zero............ 91
AC *9*; 72
ammeter................. 10
Ampère................ 9; 59
amplification 96
arc flash 137
attraction 8
battery .. 14; 15; 76; 111
binary information .. 15
BJT 95
capacitor 31; 98
CENELEC 138
chassis 126
conductance 20
conductive materials 23
conductivity 90
conductors 11
conservation................8
Coulomb............. *8*; 13
current............... 37; 44
DC *9*; *72*
diamagnetic 56
diode 11; 92
electric charge..............*8*
electric current............*9*
electric field.............. 12
electric potential 11
electric shock 17
electrical energy 14
electrical noise 18
electrical safety....... 146
electricity 7; 57; 117
electrodes 14
electrolytes 14
electromagnetism 57
electromagnets.......... 60
electron-hole............. 91

electronic devices.... 90
electrostatic force.......8
EMF......................... 16
EMI.......................... 17
energy storage........ 116
Faraday 79
fault protection........ 19
FET........................... 95
filtering 33
ground faults.......... 141
grounding................ 17
hybrid systems 132
IEC......................... 137
impurities 90
inductor 31
insulators 11; 16
integrated circuits.... 97
inverters................. 117
IoT 109; 113
junctions.................. 91
Kirchhoff 41; 42
LED 77; 94; 121
LOTO.................... 139
low-voltage.............. 19
magnetic field 14; 34; 55
magnetism............... 54
mechanical energy... 62
microprocessors 93
monitoring systems 118
MRI.......................... 64
NEC....................... 137
Ohm.................. 20; 32
ohmmeter................ 22
oscillators 97
OSHA.................... 137
overloaded circuits 141
paramagnetic........... 56

permanent magnets 56; 62
photovoltaic............. 76
power distribution ... 74
PPE........................ 136
quantization *8*
relays 64; 100
renewable energy .. 116
repulsion....................8
repulsive effect......... 56
resistance...................20
resistor 23; 31
right-hand rule.........58
robotic systems 125
safety.......................17
semiconductors . 11; 90
sensors 126
signal processing ... 112
silicium.....................90
solar cells 14
solar panels 76; 117
solar power 116
solenoid....................60
supercapacitor 99
sustainable energy . 110
switching 96
tesla 55
transformers 34; 63; 79
transistor 11; 92
transmission............73
UPS 74
voice coil 65
voltage ... 11; 15; 37; 63
voltage divider..........23
voltmeter..................15
wind power 116
zener diodes..............99

Made in the USA
Columbia, SC
28 April 2025

57230202R00096